ENERGY,
VULNERABILITY,
AND WAR

ENERGY, VULNERABILITY, AND WAR

Alternatives for America

WILSON CLARK and JAKE PAGE

W · W · NORTON & COMPANY
NEW YORK   LONDON

Copyright © 1981 by Wilson Clark and Jake Page
Published simultaneously in Canada by George J. McLeod Limited,
Toronto.
Printed in the United States of America
All Rights Reserved
First Edition
W. W. Norton & Company, Inc. 500 Fifth Avenue, New York, N.Y. 10110
W. W. Norton & Company Ltd. 25 New Street Square, London EC4A 3NT

Library of Congress Cataloging in Publication Data
Clark, Wilson.
    Energy, vulnerability, and war.
    1. United States—Energy policy.   2. United
States—Foreign relations.   3. United States—
National security.   I. Page, Jake.   II. Title.
HD9502.U52C58   1981   333.79'0973   81–9648
ISBN 0–393–01468–1   AACR2
ISBN 0–393–00059–1 (pbk.)

1 2 3 4 5 6 7 8 9 0

DEDICATION

This book is for Henry Adams, who
gave us the warning, and for the children, who will have to pick up the
pieces.

Contents

Figures and Illustrations

Acknowledgments

During the writing and preparation of this book, Susanne Page and Colleen Crossland sustained our efforts and gave happily of their time, criticism, and stewardship.

The authors appreciate the help of the many participants and contributors to the Defense and Energy Project, from which this book arose. A special note of thanks is due Bardyl R. Tirana, who inspired the study when he was director of the Defense Civil Preparedness Agency before it merged with the Federal Emergency Management Agency. James W. Kerr, project manager of the study, offered advice and rendered invaluable assistance during the preparation of this book.

Three individuals who made substantial contributions during the preparation of the Energy and Defense Project report are: David Zocchetti, Robyn Boyer, and Maurice Crommie. We acknowledge the work and ideas of the many individuals who contributed to the project, both in an advisory sense and as writers. They include the following: Jay Baldwin, Ann Bartz, Tyrone Cashman, Ann Cooper, Ralph Dieter, Deni Greene, Robert D. Lamson, James I. Lerner, John McCosker, Jan Philbin, Bennett Ramberg, Olivia Robinson, Charles Ryan, Major General Frank J. Schober, Jr., Peter Schwartz, Kenneth D. Smith, Dennis Sykes, Mary Lou Van Deventer, Leon G. Vann, Jr., Emilio E. Varanini III, and Jerry Yudelson.

In addition, the following individuals have assisted in the technical preparation of materials related to both the project and the book: Mary Christofferson, Jeanette V. Contee, Sharon Lea Gardiner, Gail L. Holt, Cheryl Imperatore, Betty Mayo, Steve Mendosa, Elizabeth Nelms, Linda Riger, Kathryn Shedlosky, and Deborah Tinitali, along with the design firm, Beveridge & Associates in Washington, D.C., especially Phil Jordan and Greg Williams.

Many individuals gave tangible assistance and ideas, including Tina Hobson and Bill Holmberg of the U.S. Department of Energy, and Louise Dunlap and the staff of the Environmental Policy Institute. Nancy Davis and others of the institute's staff were particularly helpful in disseminating information. R. J. Smith helped throughout. Jim Alexander and his associates assisted our efforts.

We appreciate the instantaneous understanding of the importance of this subject on the part of Perry Knowlton, George Brockway, and Jim Mairs, who saw to the publication of this book.

Finally, we are well aware that this book could not have been compiled without the pioneering work of a number of giants, on whose shoulders we stand. Many of them are quoted herein. We, of course, assume full responsibility for our interpretation of many original, prior works.

Introduction

This book is concerned with the current, perhaps fatal, energy disease infecting the United States and its neighbors around the world. It is an attempt to point out remedies that may arrest the progression of a condition that contributes to the frightening spectre of global war. It is a book of hope, in spite of the eminently reasonable perception that our way of life is in jeopardy.

The United States has reached a precarious state of national vulnerability, caused both by our increasing reliance on imported strategic materials and energy sources and by the less obvious danger of the centralized nature of our high-energy, technological society. The idea of vulnerability to the OPEC nations that control the heart feeding the arteries of the world's oil lifeline should come as no great surprise—similarly to those nations that control the vast reserves of strategic materials on which the machines of high technology are so critically dependent. But these obvious examples of vulnerability reveal only a part of the picture. The United States has become vulnerable to such external threats largely as a result of the creation of a highly centralized and wasteful energy system. This Trojan Horse of centralization cannot be dismembered, and its threat thus relieved, by either the creation of material stockpiles or the funding of military interventions in the Middle East.

Our system was designed to run on what was thought to

be a cheap, virtually inexhaustible energy supply: fossil fuels. When oil was selling in the early seventies for three dollars a barrel, the efficiency of the system was not an important consideration. Today, with oil prices *beginning* at ten times this level, our economy is crippled by the inefficiencies of the energy system. And our national security—dependent on energy and economic viability—is concomitantly weakened.

Our security is weakened not only by energy waste but also by the inherent centralization of the system. The more we centralize our energy systems into Brobdingnagian temples, single clusters or centers of refineries and power plants, the more, to be sure, we stick out to our adversaries a reachable chin, in the form of a small array of big and important targets. Thus we invite an enemy to believe that, quite simply and at little cost, he could cripple us. It would take a minimal attack by an adversary, or only a relatively unsophisticated strategy on the part of a band of terrorists, to bring to its knees the energy system of the United States. As we will point out, it would take the explosion of two unexceptionally large thermonuclear bombs over the United States—at a seemingly harmless altitude of some sixty miles—to eliminate a good deal of our elaborate communications and energy grids. A phenomenon called electromagnetic pulse (EMP) accompanies such nuclear blasts, making lightning seem tame. This nuclear pulse is capable of burning out the sensitive electronic equipment that controls communications and power; it would render us fundamentally mute.

Furthermore, evidence is accumulating to the effect that thinking "big" and building ever bigger installations to power the nation is leading to inefficiencies both in operations and in capital costs—thus fueling the engine of inflation. And in the meantime, as we shall show, strategies for civil defense are essentially nonexistent.

This book is based on a study commissioned by the United States Department of Defense in 1979. The study was directed by the senior author of this book for the Defense Civil

Preparedness Agency, an organization that has since been merged into the Federal Emergency Management Agency. The study, performed by the Energy and Defense Project and bearing the same title as this book, was released in early 1981. We have attempted in this volume to condense the ideas and conclusions of the study and to clarify them for a wider audience than the few hundred specialists who take note of such issuances of the federal government.

SOLUTIONS AND CHALLENGES

This is, in fact, two books. Part I attempts to explain the interrelated essence of several major national problems. Once these problems are perceived holistically, their comprehensive solution, we believe, becomes readily apparent. Part II is therefore, in a sense, a catalogue of practical strategies and technologies now in hand whereby we can make an *immediate* impact on the energy disease and, within a decade, find the problem waning, if not entirely relieved. Our national dilemma will not be resolved by vast, centralized projects oriented, like the MX missile system, toward a certain concept of defense, nor by government-subsidized nuclear and synfuel plants created to cure the energy disease. The solution lies in all of us doing our chores more assiduously, in our gathering up new and old technologies in such a way that they will deliver as much or more energy into our lives as we now have but at less cost in waste. It is a matter of doing small things—individually and as communities, municipalities, and regions—of thinking not big, and not small, but thinking *well* both about the currently available technologies and renewable energy sources and about those that will be available to us in the near future so that we can achieve maximum service, continuing economic growth, and a diminishing vulnerability.

As we shall see, there are many available technologies that can be utilized in our great industries, in our transportation systems, and even in our homes to help us use energy more effectively. In making use of these newer and better methods, we have the unprecedented opportunity of increasing the health of our economy while simultaneously decreasing the disastrous level of imported energy and material sources. In short, we can thereby solve the nation's vulnerability problem.

There are practical, democratic, even rather comfortable solutions—if we all act now, somewhat in the manner of the original American pioneers who would join in the wilderness to build a neighbor's barn or in the manner of the nineteenth-century Americans who tinkered around with felt needs and produced almost overnight, certainly within a generation, a new American presence, a new American calling: invention. Implementing the effective technologies, however, will not be as romantic a procedure as exploring space or even linking up our great cities by the Interstate Highway System (which, by the way, was originally sold to Congress as, in part, a matter of national security).

None of this can be done by the federal government alone; we are long past the time when we can rely on that organ to handle exclusively such affairs. It is something that citizens can play a direct role in—in a variety of ways and with a little wisdom on the part of the government and the business community. It can, moreover, be done with remarkable speed.

This book is no call to arms, no radical summons to a new set of basic ideals. Rather, we have attempted to indicate—with a realistic perspective on techniques, present and future—where our national interests lie. It is a tract that asks people to work in their own very best interests and altogether within the framework of the democracy we created just over two hundred years ago. We do not imagine this book is going to make anyone happy in all of its parts, for there is no ideology within the political spectrum of this nation today that confronts the curious dialectic of technology and war in a straightforward manner. We hope, in fact, that this book will generate a bit of positive anger and debate.

We also wish that the problems we point out—especially the danger, as we see it—did not exist. Unfortunately, it does not seem that the problems are misstated or the danger chimerical. We hope that the solutions we have suggested appear practical and suited to the best instincts and the noblest goals of this remarkable country.

WILSON CLARK
JAKE PAGE

Waterford, Virginia
June, 1981

Part One

THE PROBLEM

But it is a characteristic of wisdom not to do
desperate things.

—HENRY DAVID THOREAU

1. Scenario for a Short War

BAYTOWN, TEXAS—APRIL 16, 1984

As had been predicted on television by an eager young weather forecaster, a cool high-pressure zone had moved smartly to the northeast, dragging in its wake a low-lying cloud cover and the humid air from the Gulf. It was 4:00 in the morning, and there was very little happening in Baytown; most of the citizens slept.*

In the area to the northeast of town, where great fields of circular tanks and vast Gothic superstructures rose into the night, lights shone, and uniformed guards with dogs patrolled. For Sgt. Edward Flower it had been a long night—not just because of the business of patrolling the wire-fenced confines of the Exxon installation, a vast acreage of machinery and tanks, of light and dark, of the smell of petrochemicals that gave his allergies the wherewithal to create misery night after night. He was used to that. And it wasn't the return of the humid weather, which didn't help his allergies either. He had not grown up in the nation's biggest oil refinery town to worry about such things, or about the fact that for the most part, day or night, the sky looked like a bruise. On the contrary, Ed Flower thought, along with management, the environmental laws that, thank God, the government was chopping down at the knees were a sin against the

*All times in this chapter are converted to Central Standard Time.

American way. Sergeant Flower didn't give a damn if he saw
a bunch of stars in the sky. What do they have to do with
making a living? Sergeant Flower took some pride from the
fact that Baytown, his town, refined nearly 4 percent of the
nation's oil. Ambling along beside the German shepherd,
whose head was down as he menaced the night, he passed a
three-hundred-foot-high tank of good ole Baytown-refined
oil; he was proud—of Baytown, of Texas, and (except for
those clowns in Boston and New York) of America.

Ed Flower was distantly and happily aware of the squeak
of leather. He wore polished black boots. The holster for his
revolver creaked with the rhythm of his footsteps, a comfort-
ing presence. In his hand, a small black rectangular box from
which a short chromium aerial stretched toward the orange
night sky gave him the certain knowledge that he was in
direct contact with the other seventeen men hired by Exxon
to patrol this refinery. Most of these were his old buddies—
from a Baytown high school, or from the neighborhood bars.
Sergeant Flower was the sort to feel at home in the Exxon
refinery at Baytown at 4:00 in the morning.

But not tonight. Sergeant Flower had listened to Dan
Rather (a Texan, even though he was getting kind of eastern-
ized) on the TV. And Dan Rather, with his eyes darting back
and forth, had given a staccato report on the worsening situation in the Mideast. The Saudis, for the past few weeks, had
been showing signs of a collapse. There had been terrorism
directed at Saudi oil fields; the sheiks, desperately calling for
more American aid in the form of planes and weapons,
looked like a fair bid to be overrun by the Palestinians and
all the other revolutionary people who had been hired to
work in the oil fields. The president had said some strong
things, about how there wasn't any way the United States
should sit by and see a bunch of left-wingers and radicals take
over the Saudi oil fields like that loon Khomeini had done in
Iran. American interests were at stake. Hell, Sergeant
Flower had thought as he had watched the bad news

accumulate on his TV screen, American survival was at stake.

It hadn't surprised Flower at all when President Reagan, the week before, had put into effect a contingency Crisis Relocation Plan and gotten a bunch of useless people to evacuate some of the cities of the United States. (It also made the sergeant feel proud that he was considered essential personnel and had been ordered by his superior to remain in Baytown, guarding the Exxon refinery at night.)

In fact, Sergeant Flower told himself at 4:02 A.M., it probably meant some kind of war. He had never been in a war and didn't really know what a war might mean, but he had thought a lot about it in the last few weeks as things heated up in the Middle East, and the headlines were showing that the people in Pakistan, wherever the hell that was, had nuclear weapons now—thanks to the damned French who had built them a nuclear power plant—and Poland was still trying to get rid of the Russians, and the Russians were putting more tanks on the border, and the Germans—our Germans —were telling them where the hell to go. And Reagan— might have been born in Texas—had said to all those foreigners, "Okay, cut it out you guys. I mean business."

Such was the train of thought of Sgt. Edward Flower when he reached the white Ford sedan parked by the wall of the refinery's guardhouse. He opened the back door, and the menacing dog leapt obligingly into the back seat. Flower eased his bulk into the driver's seat and turned on the ignition. Nothing happened. How the hell had the battery gone dead? Damn. He turned the ignition key again with the same result. He would be late getting around to the other side of the refinery, where the precise schedule of security called for him to be—at exactly five past four. He flicked on the switch to his walkie-talkie and through the rattle of static called his next contact.

"Hey, Billy?"

"Yeah, Ed."

"Damned car won't start. I'm gonna be late. Whyn't you just go on to your next post, and I'll get over there fast as I can."

"Okay, Ed."

Sergeant Flower opened the door and stepped out into the orange murk of the refinery. Abruptly, his walkie-talkie buzzed.

"Hey, Ed?"

"Yeah."

"*My* car won't start either. What the hell is going on?"

"Just stay where you are Billy. I think we got some kind of sabotage going on here. Let the other guys know. And stay there. Tell them all to stay put."

Flower let his dog out of the car, drew his revolver, and ran to the guardhouse. In a few moments he ascertained that the telephone and the radio were both out. Flower was terrified. He went outside, called the other guards with his walkie-talkie, and told them to be on the alert for a terrorist attack.

At that moment there was an overwhelming flash of orange-yellow light overhead. Before Sergeant Flower's eyes could register this phenomenon on the nearby cells of his brain, he and his dog were gone, as were all the refineries in Baytown, Texas. It was 4:13 in the morning.

There is no way available to the human mind to comprehend the simultaneity of events that occurred in the wee hours of April 16, 1984, in the United States. The most awe-inspiring events all took place within fractions of a second, but the details take several minutes to read (and perhaps years to absorb as a matter of intellectual understanding).

It was a desperate but very shrewd war. . . . The Soviet Union had determined a number of things. Its own position was weakened in the world by the resurgence of nationalism among its satellites, the proliferation of nuclear capability among unpredictable non-client states in the Indian subcontinent, and the difficulties it was having producing food. But its chief concern was that the United States, reacting rationally if slowly to the growing threat posed by the disintegration of the Saudi regime and the terrorist sabotage of the Middle East oil fields, was plainly preparing to take over in Saudi Arabia, despite the obvious risks.

The Kremlin decision to make a surgical strike on the United States was designed to accomplish two goals with frightening speed: first, to destroy the communication links in the nation, except for those between the White House and the Soviet Union; second, to destroy the bulk of America's oil refineries, making an early economic recovery impossible. Europe was not spared in this doomsday strategy. In that theater, key military installations housing nuclear missiles, submarines, and bombers were targeted, leaving an economic infrastructure but eliminating any European capacity to defend against the Russian tide that would soon come— first soldiers, then bureaucrats. Europe would be Russian and so would what was left of England.

In the United States, all the carefully laid plans for critical communications in the event of war were rendered useless in the eight minutes it took to explode the two submarine-launched ballistic missiles about sixty-five miles above Tennessee and Nevada. A carefully timed barrage of submarine missiles was targeted on coastal oil refineries. The Kremlin hoped that its dramatic strike would cripple the United States in such a way that it would have nothing to do but lick its wounds and understand, in the long course of rebuilding itself, that it was now a minor, even insignificant nation. Such was the thinking of the Kremlin planners, and they believed their plan would work, even though it was risky; one could not rule out a maniacal and useless retaliatory strike.

The Russians knew what most average Americans seemed to be unaware of: that there was no American capacity for shooting down the Russian missiles—no anti-ballistic missile system—and that there still, amazingly, remained vast holes in the American radar coverage of its territory. So they could be fairly certain that with a little bit of the surprise element remaining, they could reach their targets.

The war started as a result of the use of theater nuclear weapons, as had been predicted in the 1970s by an East German scientist, Robert Havemann. Havemann wound up in an East German prison for "leaking" Soviet fears about a U.S. first strike, based on the use of "theater" nuclear weapons. He pointed out that the NATO countries' acceptance of U.S. cruise missiles and Pershing II missiles would have a stunning effect on the Soviet leadership, who in fact likened this development (of the early eighties) to the Kennedy-Khrushchev confrontation over Cuban-based Soviet missiles in 1962. Since the Pershing IIs and the cruise missiles were operated by U.S. personnel on foreign territory, the Soviets never accepted the West's definition that the nuclear warheads were solely for the protection of Western Europe.

They were, instead, perceived as a principal *strategic* nuclear arsenal, capable of delivering a powerful first strike deep into Russian territory. In fact, the weapons could reach some five hundred miles farther into Russian territory than the low-penetration F-111 bombers available to the United States. In addition, they could be exploded on the hardened Western Russian missile sites and political centers in about five to seven minutes. The Kremlin was keenly aware that the elaborate civil defense shelters built for the highest echelons of Moscow's elite would be useless—Moscow would cease to exist five minutes after the button was pressed by an American missile crew on the outskirts of Hamburg.

The war started as a result of a nuclear detonation about 150 miles directly west of Mosjöen, Norway, just below the Arctic Circle. An American guided-missile cruiser, the *Intrepid*, had indeed launched a small nuclear-tipped, antimissile missile. The decision was made by a forty-one-year-old Navy veteran, Captain Harry Barnes, who had acquired the tiny nuclear weapons for his frigate only seven months before. His mission was to protect an important convoy carrying critical rapid deployment force supplies to the Norwegian port at Trondheim, for secret transfer to the NATO bases in that country. At 2:48 A.M. on the sixteenth of April, 1984, Captain Barnes received a top-secret communication that two of the new Soviet Oscar class titanium-hulled attack submarines had been detected about fifty miles northeast of his position. Barnes put his crew on full alert and received Pentagon authorization through the NCCS (Navy Command and Control System) satellite to arm his ship's anti-cruise missile nuclear warheads.

At 3:45 A.M., the *Intrepid* launched three of these specialized, sleek twenty-foot-long missiles. History has not recorded the details of the moment. Presumably, Captain Barnes responded to a launch of cruise missiles by one of the Soviet Oscars, aimed at the ships in the convoy he was guarding. Regardless of the incident itself, the detonation of the *Intrepid*'s nuclear missiles in that remote sector of the North Atlantic was interpreted by the Kremlin was the first, irrevocable stage in European hostilities. World War III had begun, and the Soviet reaction was swift—faster, in fact, than the West had ever believed possible. Fifty Backfire bombers assigned to the Soviet Northern Fleet were airborne within minutes, and nuclear-armed cruise missiles were targeted on the NATO convoy, as well as other vessels and ground targets in Scandinavia. A myriad of preprogrammed weapons systems were put on the alert and armed.

Within minutes, the Soviet high command decided to act swiftly and, among other things, knock out the vital military communications network (including satellites) of the United States and her NATO allies. This was accomplished by the simultaneous explosions of high-altitude nuclear warheads over the United States and Europe. This carefully timed attack wiped out much of the critical military communications network necessary to program the elements of war.

What happened to the United States as a result of this train of events took place too fast for most people to be able to understand, much less reconstruct. Sergeant Flower, for example, was subject to an immediate effect of the nuclear explosion. He found that the ignition to his car would not work when, a few minutes past four o'clock, he attempted to start it. This was the result of a precursor to the SS-18 missile attack, the two relatively small Russian bombs (in the kiloton range) that exploded sixty-five miles above the United States. Within a minifraction of a second, most of the electronic systems of the United States were rendered useless.

High-altitude detonations of relatively small nuclear weapons create a microsecond burst of energy, called electromagnetic pulse, or EMP. It is similar to lightning but exhibits a rise in voltage a hundred times as fast. Conventional electric power equipment designed to protect against lightning is helpless in such a pulse. Further, solid-state electronic equipment, communications systems, and the electric power grid are exceptionally vulnerable to EMP. Transistors burn out permanently. Electrical grid systems (sensing a considerable problem) shut down. Short-antenna walkie-talkies like Sergeant Flower's would not be affected, but the electronic ignition system of his Ford would be out of business.

The same is true for the ignition systems of the trucks that take SAC bomber crews to their B-52s. Modern military weapons are regularly subjected to tests of manmade EMP to be sure that they are immune. The SAC bombers still operate on vacuum tubes, not the chips that are so prone to EMP shutdowns, but seconds after it became apparent that the nation was being attacked, various flight crews scrambling into their standard Chevrolet trucks in a headlong rush to their B-52s, found the electronic ignitions inoperative. They ran, cursing, the half mile to the big antiquated planes, but by then, of course, it was too late.

Minutes later the SS-18s struck and the U.S. economy was laid waste.

In Baytown, there was an explosive blast, driving air away from the site of the explosion—producing what is called static overpressure (a sudden change in air pressure that can crush objects) and high winds, called dynamic pressure, that can move objects suddenly or knock them down. Most of the large buildings in Baytown collapsed because of static overpressure; it was dynamic pressure that killed most of the people in the vicinity by blowing them into other objects. Walls of two-story houses four miles from the blast were pushed over by pressures of about 180 tons; people were blown away by winds of 160 miles per hour. All of this in seconds.

About one-third of the energy in the explosion over Baytown was given off as heat, a blinding flash that lasted maybe two seconds—first temporarily blinding everyone within thirteen miles of the blast. Most of the people of Baytown who had not vacated were asleep, so they were not blinded, but thousands who were up and about within five miles of the flash were severely burned. Since heat travels at the speed of light, Sergeant Flower and his guard dog were blinded, burned, and blown away in that order, all before he could possibly know what had happened, never to feel the effects of the ionizing radiation that would spread throughout the region, particularly in the direction of the winds that were taking the clear cold front to the northeast of Baytown and would render Sergeant Flower's suburban neighbors ill and, in many cases, dead in the days to come.

Again, within seconds of the blast, Baytown was a raging inferno of fire, caused by the blast's heat and the chain-letter effect of petroleum installations collapsing and catching on fire with incendiary power. The sky over Baytown was red; the vast acres of refineries, pipes, and tanks were destroyed —leaking oil into nearby waterways that were themselves burned, incinerating barges and tankers, and leaking poisons from the devastated petrochemical plants into the water and into the surrounding hell that had been Baytown.

Except for the people dying of burns, except for the people who would die later from radiation fallout, and except for the fires consuming the wreckage, it was all over in a few seconds.

And not, of course, only in Baytown.

The Soviets had targeted the major oil refining areas of the United States, mostly clustered in a few places. The footprints, that is, the regions subjected to the effects of this nuclear attack, numbered ten: one in Northern California; one in the Los Angeles area; two running from Indiana to west of Chicago; one south of Kansas City; four in areas overlapping from mid-Texas to Alabama; and one in the middle Atlantic, covering a region from south of Washington, D.C., well to the north of New York City. In all, nearly two-thirds of the U.S. oil refining capacity was eliminated in one well-timed series of simultaneous strikes. The purpose, an economic one, had been achieved. Scenes like Baytown occurred in virtually every large industrial area of the United States. Though people were not the target, more than five million died in the first moments of the attack.

Philadelphia took two direct hits, both near the Schuylkill River in downtown. The Liberty Bell was vaporized. Of the 155,000 people within a two-mile radius, 135,000 were instantaneously killed. Of the 785,000 people within five miles of the twin blasts, an area stretching east into Camden, New Jersey, and west to the city's limits, 410,000 people died. Of the four electric power plants in the city, two were badly damaged, one moderately damaged, and a fourth was left largely intact. But demand for electricity was not great in the following few days. The Philadelphia airport was wiped out, a mangled fiery ruin. Ships were burning in the harbor channel, grounded and sunk in the narrow reach near the naval shipyard. Railroad tracks were severed, cutting off rail transport from Philadelphia to anywhere, and highways—major northwest-southwest arteries—and bridges had ceased to exist. Eight hospitals were destroyed in the vicinity of the blast; burn and blast victims were prevented by the rubble and fire from reaching the hospitals beyond the immediate effects of the attack. Even so, these and outlying medical facilities soon became overtaxed, and thousands of people died agonizingly of burns.

The drydocks of the U.S. Naval Shipyard were a heap of rubble; nothing was left of the Defense Supply Agency complex. There was no more University of Pennsylvania, no more Independence Hall.

Seconds after the nuclear detonation in the North Atlantic, the president was awakened by his military aide. He was quickly shuttled aboard a waiting Marine Corps helicopter for the short flight to Andrews Air Force Base, where he boarded the waiting, specially equipped military version of the Boeing 747. This plane, known as the *Kneecap* plane (military jargon for the acronym of National Emergency Airborne Command Post), was equipped and outfitted with EMP-resistant communication circuits. It was the only secure communications center in the event of nuclear war between the superpowers. With the president was his ever-present military aide, who carried the authorization codes necessary to launch virtually the entire U.S. nuclear arsenal.

At this point, there were no civilian communications of any significant scale left in America, unless one considered CB radios. Radio, television, and all of the other media by which Americans could have learned what had happened while they slept were out. But information poured into the crowded compartments on the *Kneecap* plane. It also poured sporadically into some seventy other terminals located in concrete bunkers in the Shenandoah Mountains and elsewhere, places where the leaders of the United States might foregather in the event of a crisis. The president, however, had chosen in the days previous to the attack to stay at the obvious helm of the nation, the White House, in spite of the vital decisions he had had to make.

Within moments of the time the president stepped on the plane, even as it was gaining speed down the runway, a young brigadier general, white-faced and trembling, lurched against the president and thrust into his hands a paper with the familiar capital letter of the Hot Line telex. Around him, military personnel were talking with remarkable calm into their microphones, the maps and printouts gleaming green on their screens. At the top of the paper, the president noted, was the beginnings of a description of a move on a chess board. "QB to . . . ."

There was a new chess game now. The communication was from STAVKA.

STAVKA, the president well knew, was the mechanism, in place for thirty years in the Soviet Union, which would take over all power in the Soviet Union if a war broke out. It was made up of many of the leaders the president had met during his trip to Moscow but also included others, whom no one in the United States had ever laid eyes on. It was a sleek, well-organized, transformed version of the Soviet's Main Military Council, of which the premier was merely one member. The president read STAVKA's message: MR. PRESIDENT: THE UNION OF SOVIET SOCIALIST REPUBLICS HAS ELIMI-NATED INTERNAL COMMUNICATIONS AMONG YOUR PEOPLE. WE HAVE ALSO ELIMINATED YOUR ENERGY CAPACITY. ALL OF YOUR EUROPEAN ALLIES AND JAPAN ARE DISARMED. RE-TALIATION WILL BE OF NO AVAIL. WE REPEAT. NO AVAIL. IT WILL LEAD ONLY TO THE END OF YOUR PEOPLE. INSTEAD, YOU MAY BEGIN TO PLAN THE REBUILDING OF YOUR NATION, A TASK WITH WHICH THE UNION OF SOVIET SOCIALIST REPUBLICS WILL ASSIST. It was signed Ustinov. That was Marshall D. E. Ustinov, the minister of defense. The months of talking to the premier of the USSR had been wasted.

The country's oil supplies and capability were cut by two thirds, at least. It is clearly the end of America, as we know it, thought the president. It is clearly an end to California and the ranch and . . . a dozen images flitted through his head.

Response. Response. We have been blown away, the president knew. If they have taken out our energy systems, we might try the same. Yet, from many briefings of the past months, the president knew well that a similar response would be less effective, the conditions of energy being different in the Soviet Union. It was 4:38 A.M. Reports of devastation flooded in, filling in the map with green lights winking at him. No, he had two choices to mull over. He could retaliate in full. Or he could wait for the other shoe to drop. He could unleash a force of missiles and destroy a similar component of Russian society, in which case the other shoe would certainly drop on the rest of the U.S. capacity to function, or he could wait. He chose to wait.

He was right. There was but one superpower left in the world. The United States was out of business. There was no purpose to another Russian strike. And a retaliatory strike was of no use. America had decades of rebuilding to do. The president knew this, knew that the Soviets had accomplished their goal, that the United States was basically an economic basket case and could not risk the rest of its potential on extending the war. The president knew that, by 4:39, the Third World War was over. The Soviets had won. He knew that they now owned Europe—and that, as of that moment, he had to take unto himself all powers, remove from the books all civil and public liberties, take command of all the forces of government, as would a tyrant, in order to rebuild a nation that, in a few short seconds, had had its life blood sucked out of it like that of a fly by a spider. He knew it would take a generation to rebuild, if not longer.

The president waited for the second shoe to drop, and it did not. At 4:50 on April 16, 1984, the president sued for peace.

*　　*　　*

The foregoing is a dramatization of an analysis made by the Congressional Office of Technology Assessment at the request of the Senate Foreign Relations Committee, in order to assess the effects of a nuclear war on the United States. Our scenario is based on one of several offered in the report, *The Effects of Nuclear War;* the others were the more traditional accounts of larger-scale attacks. This one, the authors wrote, "is representative of a kind of nuclear attack that, as far as we know, has not been studied elsewhere in recent years—a limited attack on economic targets." It is, one might say, an energy war and it is altogether possible that we will see such an energy war erupting.

Indeed, the Iraq-Iran conflict has been just such a war. Most primary targets in the early months of the war were energy targets, ranging from refineries and key oil fields to an Iraqi nuclear research center. The only missing ingredient in this Middle East duel was nuclear-tipped warheads. That this kind of war might become a trend was suggested by a recent report in the *Washington Post.* Commenting on each side's declared intent to bring the other to its knees economically, a Western diplomat was quoted as saying, "If this is going to be the pattern in the Middle East, then it is an ominous prospect for those of us so dependent on imported oil. Combatants in the Middle East have always scrupulously avoided hitting each other's economic installations. There obviously are new ground rules on this score."

It is self-evident that such outbreaks in the Middle East spotlight our own vulnerability, since critical oil supplies must pass through the narrow Straits of Hormuz. Yet there is another form of vulnerability from which the United States suffers, and the scenario above highlights it: our life-blood— oil—is one of the most concentrated and centralized systems in a highly centralized industrial society, as is our electrical generation system. Both are subject to attack by terrorists or an adversary nation. Indeed, it is our contention that as these systems become more rather than less centralized, they virtually *invite* attack.

Would the United States be able to recover from such an

attack? It is useful to review the findings of the Office of Technology Assessment in this regard.

* * *

RECOVERY

T. S. Eliot was right about April. To continue our scenario, by midday of April 16, 1984, most Americans knew that the central assumption of their lives—that there can be no nuclear war—had been obliterated. They did not know the extent of the damage, and it was some time before they knew that there would be no more attacks. Until this knowledge was broadly appreciated, the undamaged regions of the nation withheld aid to the attacked areas.

The immediate tasks in these savaged regions were care of the injured, burial of the dead, searching for other victims, and fighting the fires. Refineries, tanks, and petrochemical factories burned out of control for days, threatening yet other installations. Unseen doses of radioactivity had rendered much of the spring planting of America's farms useless, though few were aware of it at the time. April *is* the cruelest month.

Most people within five miles of the blasts who had not been killed immediately were dying of burns; there was little help for them. In peacetime, the United States had had sufficient facilities to take care of only a few thousand burn cases at any time; some of these facilities had been destroyed; others were isolated by the large-scale transportation breakdown caused by the attack. Most third-degree-burn victims died.

By the end of the first week, people knew for certain that the war was over. Those who had evacuated the cities before and after the attack began to return; the task of piecing together an economy began.

With only one-third of its oil remaining, the government immediately imposed Draconian rationing. Petroleum products were reserved for critical industries and services—military and police, agriculture-related equipment, railroads, and firefighting equipment. Most of the country was past the season when home heating oil was required, but in several months a frigid winter would be on its way. The massive interconnected spider web of production, services, and jobs associated with the automobile was ruined, and for good. Millions of people—from steelworkers to gas-station owners to fast-food restauranteurs—were permanently out of work.

Along with the loss of two-thirds of oil refining capacity, about half of American natural gas supplies were severely disrupted. Seventy-four percent of domestically produced natural gas had come from the devastated Gulf Coast area, and a great deal of all natural gas traveled through the area in pipelines, many of which had been severed. Only a few harbors equipped to receive imported fuel were still in operation, notably Norfolk and Seattle.

Most of the initial 5 million deaths—except later radiation deaths—were over by the first month. With millions of people out of work, demand for aid and unemployment compensation rose just when tax revenues were plummeting. Mass transit—within cities and between cities—was inadequate, and people simply did not travel; they stayed home and, when they could, bartered for goods that were undergoing staggering inflation.

Most vital services, including banking and finance, were taken over by the government, but the array of useless factories (whose products were no longer needed) and service industries with unemployed workers led to a situation in many ways resembling the Depression of the 1930s—and in many ways worse. Most workers and most equipment had of course survived, but it took years—decades—for an economic recovery to take place.

The restored nation was an entirely different kind of place. Energy conservation of all kinds was imposed by regulation: all new houses were better insulated than before, most were fueled by alternative energy sources. Mass transit, railroads, ships, and bicycles replaced automobiles, trucks, and air travel. Agriculture survived but with notable changes—particularly, the raising of livestock, the most energy-intensive agricultural practice, was sharply curtailed, and meat became an expensive luxury that few enjoyed.

Public health both declined and improved. Mental illness was epidemic in the first years of recovery. The long-term effects of radiation in the afflicted areas claimed many in the years to come, as did the toxic petrochemicals that had polluted large areas of land and many bodies of water. For years people lived in substandard housing, and illness increased. But a few of the patterns of life now forced on Americans temporarily promoted health—for example, there were fewer auto accidents and the people benefitted from the exercise of walking and bicycling.

* * *

No one, of course, can predict exactly what shape that new nation would take, but the Office of Technology Assessment report says, with logic and with an irony unusual in a government report, that it "would tend to apply the lessons of the past to future policy by seeking to reduce the vulnerabilities to the last attack . . . many people say that the United States would be better off if it were less dependent on cars and petroleum. While changing to new patterns of living via nuclear attack would minimize political problems of deciding to change, it would maximize the difficulties of transition."

THE LIKELIHOOD

In our dramatization, we followed closely the details of the OTA study—from the attack through the recovery period—because, among other things, the scenario illustrates two important points. One is a new optimism among certain planners. It suggests that a limited nuclear attack is feasible: that one side, severely crippled, might indeed sue for peace; that recovery, though painful, would occur, and a more or less recognizable nation might emerge in a decade or two.

This seems highly unlikely to us. After the initial knowledge of the attack, the assumption would almost certainly be that more was coming and the president would almost certainly retaliate.

Statesman Charles W. Yost discusses an attack on U.S. missiles in *The Bulletin of the Atomic Scientists.* Those who maintain that "an American President would not dare retaliate . . . for fear this would provoke a Soviet second strike on our citites . . . misread the psychology of any American president I have ever known. After all, what were the other two legs of the triad (submarines and bombers) constructed for? Could any President passively ignore, fail to respond to, a massive attack upon us? Did not Truman, even when we were winning, feel he had no choice but to use the atom bomb?"

In fact, we would be obliged to retaliate.

Just what form the retaliation would take is unpredictable. It could be an attack in kind on Russian energy systems, though they are more dispersed than ours and an attack such as described above would be somewhat less damaging. It could be an attack on Russian missile silos, though they might by then be assumed to be empty and the missiles on their way. Or it could be an attack on Russian population centers. In any case, in the event of a limited attack as described in this chapter, the Third World War would probably go on longer—maybe another hour or two—and both powers would be devastated in ways that are fundamentally unpredictable.

A truly limited nuclear war is a remote possibility. Yet many military planners are beginning to believe such a war would be winnable—a perverse kind of optimism.

The other point the OTA study dramatizes is that of our

vulnerability. Presenting such a small, centralized array of compact targets as our energy system might lead an adversary to believe that a limited attack could succeed. Regardless of the size of the conflict, the vulnerability of the energy system is itself a vital and frequently overlooked strategic issue.

2. The Present Danger of Energy Wars

OUR FICTIONAL account of a nuclear war—triggered in large part by mounting international pressures on energy resources—need not occur, if nations act now to avoid the escalation of the arms race. However, current trends indicate that the race is not slackening, but is rapidly increasing in its global dimensions.

GLOBAL WEAPONRY

In 1980, Frank Barnaby, director of the Stockholm International Peace Research Institute, reviewed the world arsenal, pointing out in the 1980 *SIPRI Yearbook* that military spending had increased fourfold since World War II, reaching more than $500 billion—six percent of the world's total gross product. Of this, $120 billion is for weapons; one-fourth of those weapons are exported. It was exported weapons that formed the bulk of the "weapons used in the 130 or so wars which have taken place in the Third World since World War II—wars which have killed some 30 million people."

Selling weapons is a lucrative business, and the United States is the largest entrepreneur, supplying 45 percent of the world's exported weapons. The Soviet Union supplies 27 percent; France, 10 percent; and Great Britain, 5 percent. A number of Third World nations are beginning to get into the act—Israel, South Africa, Brazil, Argentina, and India.

The great bulk of weapons produced are conventional ones, which are sufficient to make plenty of trouble around the world. But it is unquestionably nuclear weapons that pose the most serious threat. Table 1 shows the world inventory of delivery systems—either missiles or planes—as of 1980.

Keep in mind that many, in fact most, of these delivery systems can carry numerous warheads (that is, bombs)—each of which, once released, can head off by virtue of an internal guidance system to a prearranged target. A single Russian SS-18 missile can carry eight warheads. Keep in mind, also, that what Russians and Americans mean by the word "strategic" as applied to weapons is that kind of delivery system by which one of the two countries can nuke the other. Systems by which we could nuke Cuba, or Russia nuke Europe, or vice versa, are not considered strategic in the jargon.

For the past several years, the Soviet Union has increased its production of nuclear weapons and is reaching parity with the United States in terms of intercontinental power. In terms of actual megatonnage, the Soviet Union is somewhat ahead. In a sense, Table 1 is misleading, for it shows the USSR with 135 strategic bombers and 805 intermediate-range

Table 1. Nuclear Delivery Systems

	US	USSR	China	UK	France
Intercontinental Ballistic Missiles (ICBM)	1054	1447			
Medium-Range Ballistic Missiles (MRBM)		600	80		18
Submarine-Launched Ballistic Missiles (SLBM)	656	625		64	64
Strategic Bombers	373	135			
Intermediate-Range Bombers	68	805	150		36

bombers. Among the latter are several hundred Backfire bombers; whether or not they are strategic is a question of some uncertainty. In his exhaustive survey *U.S.-Soviet Military Balance,* John M. Collins points out that Backfires could not make low altitude (thus fuel-rich) flights and penetrate the continental United States. In high-altitude flights, he says, they could reach the Pacific Northwest. On the other hand, if the crew and the plane were considered expendable, this "country would be completely exposed." The Soviets' other option is to bomb the United States and land the Backfires in some friendly Caribbean or Central American nation, Cuba being the obvious choice. In his discussion of the Backfire in *Weapons,* Russell Warren Howe draws attention to the fact that as far back as 1977 *Jane's All the World's Aircraft* accorded the Soviet Union an absolute lead in aviation, in part because the Backfire bomber was clearly a strategic weapon (even though it was excluded from the SALT agreement).

In a 1980 review of NATO and the Warsaw pact, the *Economist* noted, "For several years Russia had outreached the United States in most measures of nuclear strength—megatons of explosive power [1 megaton is equal to 1 million tons of TNT], numbers of missiles, and the total weight that can be lifted to the target. Only in numbers of warheads has the United States remained ahead. But even this last American advantage is rapidly disappearing as the Russians deploy large numbers of independently targetable re-entry vehicles on their big new missiles. The raw warhead totals do not tell the whole tale anyway. A much higher percentage of America's warheads are carried by manned bombers and submarine-launched missiles. The bombers have a much smaller chance of getting through than missiles do, and the submarine missiles are not only much less accurate than the land-based ones—not accurate enough to destroy the other side's missile silos—but also less readily usable (only about half the American missile submarine fleet is at sea and ready for action at any given time)."

Arsenals are being improved, as well as added to, on both sides. The guidance system of the U.S. Missileman III ICBMs is being improved "to decrease the circular error probability (CEP) from about 350 meters to about 200 meters," reports Barnaby. That means these missiles would have a probability of 50 percent of knocking out a Soviet missile in a silo with one shot, 90 percent with two shots. The proposed MX missiles, guided by lasers or radar, will have a CEP of less than 100 meters. The Soviets are not lagging far behind in the accuracy department. The Russian SS-18 missiles now have a CEP of 500 meters, and this will soon decrease to 250. The situation is much the same for the other missile systems, those launched from submarines. The U.S. SLBMs account for 70 percent of our nuclear weapons—there are 5,120 independently targetable warheads under the seas. Before long they will achieve CEPs of less than one hundred meters.

It is the nearly unbelievable precision of these new weapons and improved old weapons that makes a war more likely, because military planners might conclude that a limited first strike—a limited nuclear war—is fightable and winnable. Consider what has come to be called Eurostrategic weapons: nuclear weapons that are pointed at or located in Europe. Pershing II missiles, planned for deployment in several NATO countries, could reach Moscow from West Germany; ground-launched cruise missiles, also planned for European deployment, could reach even farther into Russia. As Barnaby points out, "It is difficult to see why the reaction to the explosion of nuclear warheads carried by shorter-range missiles would be different from that of the explosion of warheads carried by strategic missiles."

NUCLEAR EFFECTS

It is chilling but necessary to review the effects of the detonation of a nuclear bomb.

The blast effect of the explosion of a one-megaton bomb

at eight thousand feet above the Earth's surface would level reinforced concrete structures in an area of nearly a one-mile radius. Within nearly a six-mile radius, walls of typical steel-frame buildings would be blown away and residences severely damaged. Winds would be severe enough to kill anyone out in the open. Even eleven miles away, structures would be damaged, and people would be endangered by flying glass and debris.

But the blast effect is only part of what would happen if a nuclear bomb exploded. The heat from the explosion accounts for about a third of the bomb's energy; it precedes the blast wave by a few seconds. The explosion described above would cause flash-blindness as far as fifty-three miles away on a clear night. It would cause third-degree burns up to five miles away. Third-degree burns over one-fourth of the body will cause severe shock and probably death if specialized medical help is not provided promptly. As the OTA report points out, the United States "has facilities to treat 1,000 or 2,000 severe burn cases; a single nuclear weapon could cause more than 10,000." In 1980, the American Medical Association, having stared into this particular abyss by analyzing the potential for medical services in the greater Boston area after a nuclear attack on that city (there would be virtually none), concluded that there is no cure to this sort of thing—except prevention. The thermal radiation, in addition to burning people, would cause firestorms such as those at Hiroshima with a resulting widespread loss of life.

The third and most long-lasting effect of the detonation of a nuclear bomb is radioactive fallout, which is the term for radioactive particles that are created by the irradiation of materials swept up into the nuclear mushroom cloud. Some of the fallout would immediately blanket an area within a radius of ten miles from the blast center, the rest going into the atmosphere, to fall to the ground later. Within a week, high radiation levels, capable of causing death and serious injury, would extend up to two hundred miles from the blast center. Some of the more toxic materials would be somewhat neutralized within a period of days and weeks, but many of the radioactive materials would remain toxic for lengthy periods, increasing the incidence of cancer for generations. Certain kinds of plants and animals—for example, pine trees —are more sensitive to the effects of radioactivity than others. Soil microorganisms, the basis of the entire food chain, are even more fragile. Besides causing damage to humans, to buildings, and to all other material substance, a nuclear explosion would cause devastating ecological havoc. It would be a long time before any survivors of such a blast would hear the song of a bird.

Another effect of a nuclear explosion, as we saw in chapter 1, is the electromagnetic pulse (EMP), resulting from secondary reactions when gamma rays are absorbed in the air or ground. EMP effects do not hurt people, but they can render a complex, electronic society dysfunctional. High-altitude detonations of relatively small nuclear weapons (even in the kiloton range) create this microsecond burst of energy, similar to a bolt of lightning except that the rise in voltage is a hundred times as fast. Conventional electric power systems designed to be impervious to lightning are helpless in the case of EMP. Solid-state electronic equipment, modern communication systems, and the electric power grid are exceptionally vulnerable. To protect against it, you need special surge arresters, bypass devices, hardened components, and other equipment; presently, only substantial military communications and power systems are so protected. It has been estimated that one small nuclear explosion high in the atmosphere over Kansas City would create EMP sufficient to eliminate most civilian communication systems and much transportation; the blast would perhaps cause even the loss of nuclear reactor control over the entire United States, except for a small fringe of the Pacific Northwest Coast.

There is little the United States could do in defense

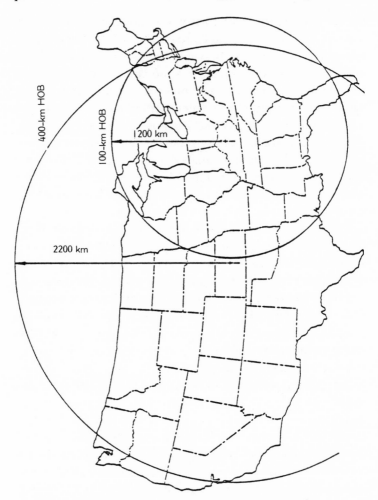

Figure 1. Area of Coverage of EMP from High Altitude
Detonations (HOB: Height of Burst)

against even so simple an attack as a couple of EMP-producing high-atmosphere explosions. In the early 1970s both the Soviet Union and the United States agreed in SALT I to limit anti-ballistic missile sites to one apiece; the United States, as John M. Collins put it, "dispensed with that prerogative." It is common knowledge among ecologists that a single crop—a monoculture—is far more vulnerable than a diverse natural system would be to damage from a mutant disease organism or a drastic change in the weather. Modern industrialized nations such as ours, dependent on centralized systems of energy, production, and communication, are like a monoculture. It would take very little to put us out of business.

The ground rules that the authors of the OTA study set themselves for the imagined scenario of a limited nuclear war, which we dramatized in chapter 1, were simply to ask themselves what the Russians might do if they wanted to fight a war and limit themselves to only ten missiles, each with eight warheads. They concluded that the Russians would attack the highly centralized lifelines of the American energy system. Just like a mutant virus in a cornfield.

WORLD WAR II

There is nothing new about targeting energy facilities during times of war. We mentioned the Iraq-Iran war of 1980–81 (and we shall return to it later), but energy systems were also targets in the Korean War and World War II. The examples of Germany and Japan in World War II offer a clear-cut demonstration of the disadvantages of a centralized system of energy production in a time of war. Electric power was the most vital part of the German energy system. By 1944 about 80 percent of all the Reich's industrial motive power was derived from electrical sources. The chief fuel for generating electricity was coal—accounting for 80 percent of both public and private electrical power.

At the beginning of the war there were 8,257 generating

stations in greater Germany. Most of them were very small; more than 80 percent of all electricity was generated by only 416 stations. Though generating stations were located throughout the country, there were five main concentrations of generating capacity, each dominated by one or more large public plants and a number of large private plants. Early in the 1930s the Germans had begun connecting the generating stations by means of high-tension transmission lines, and, beginning with the war, most of the large private power stations having a surplus of power were tied into the public utility network.

The existence of the national grid caused the Allies, early in the war, to question the vulnerability of the electric power system. It was felt, according to a 1947 report by the U.S. Strategic Bombing Survey, that the "mobility of electric power, except under limited conditions . . . would permit the Germans to spread the loss at any point throughout the region attacked, and probably throughout Germany."

The Germans themselves were not so sanguine. Albert Speer, the German minister for armament and war production, commented, "I think that attacks on power stations, if concentrated, will undoubtedly have the swiftest effect; certainly more quickly than attacks against steelworks, for the high quality steel industry, especially electro-steel, as well as the whole production of finished goods and public life, are dependent upon the supply of electric power . . . the destruction of all industry can be achieved with less effort via power plants." Agreeing with Speer, Hermann Goering, commander of German air forces, elaborated, "We were very much afraid of an attack on German power plants. We had ourselves contemplated such an attack in which we were to destroy power plants in Russia."

Throughout 1943 and into 1944 there was some loss of German power plants due to Allied action. But beginning in the late spring, as figure 2 demonstrates graphically, the Allies began to fulfill the Reichsmarshall's fears all too well.

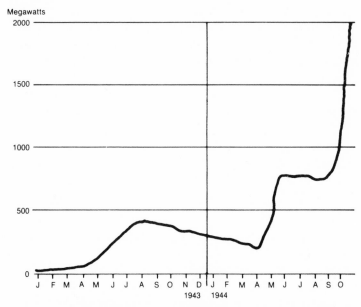

Figure 2. Air Raid Damage to German Power Plants (Damage in Megawatts of Capacity)

After the war, the chief electrical engineer for Germany's largest utility, RWE, stated, "The war would have been finished two years sooner if you [the Allies] had concentrated on the bombing of power plants earlier. . . . Your attacks on our power plants came too late. This job should have been done in 1942. Without our public utility plants we could not have run our factories and produced war materials. You would have won the war and would not have had to destroy our towns. Therefore, we would be in a much better condition to support ourselves. I know the next time you will do better."

The story is much the same for another critical component of German industry, the production of synthetic fuels. By 1941, in the Reich and its occupied territories, 150,000 vehicles ran on "producer" gas made from coke, anthracite,

Barrels Month⁻¹ x 10⁻⁶

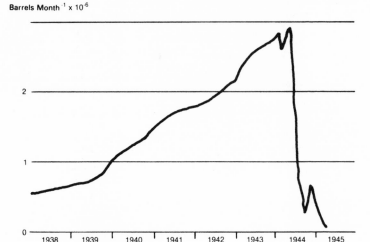

Figure 3. Air Raid Damage to German Synthetic Fuels Plants (Damage in Lost Production: Barrels/Month)

charcoal, coal, peat, and other fossil sources. By early 1944, synthetic oil production accounted for more than half of the entire German oil supply; it provided three major products—aviation gasoline, motor gasoline, and diesel fuel. By that time, more than four-fifths of the large vehicles were converted. German filling stations were established to dispense wood chips and alternative fuels. To maximize usage, the Reich granted subsidies for vehicle conversion, ranging from four hundred to one thousand Reichsmarks per vehicle.

All in all, by the end of the war there were twenty-seven plants engaged in producing synthetic fuels—concentrated, for the most part, near the major coal mines in the Ruhr Valley and thus highly susceptible to attack. But the Allies' prime targets were initially strategic military facilities. In this case too, the Allies failed to take advantage of Germany's energy vulnerability until very late in the war. (See figure 3.) When the Allies did destroy Germany's main synthetic fuel

Percent

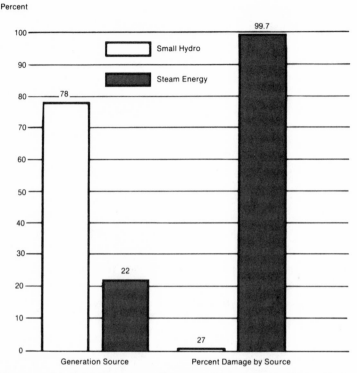

Figure 4. Air Raid Damage to Japanese Power Plants (Sources of Generation and Percentage of Damage)

plants under the bombing attacks, synfuel production dropped to virtually zero from nearly 3 million barrels a month. And, with the destruction of the main electric power plants, the German war economy was essentially incapacitated and the war, for all intents and purposes, was over in Europe.

As figure 4 shows, the situation in Japan was quite the opposite. Electricity generation in the peak war year on the home islands was 38.4 billion kilowatt-hours; of that, 78 percent was generated by water power from small hydroelectric plants. The remainder was supplied mostly by antiquated coal-fired steam plants. The total air raid damage to the Japanese utility grid consisted of bombing hits on thirty-five power plants. Indeed, according to the U.S. Strategic Bombing Survey (Pacific), the electric power system was never a primary strategic target because most of the power facilities of Japan were "so numerous, small, and inaccessible that their destruction would have been impractical, if not impossible."

FROM COLD WAR TO MIDDLE EAST

In the early years of the Cold War, during the Berlin crisis of 1948–49, a recently declassified Joint Outline War Plan called for a potential bomber strike against the Soviet Union with 150 nuclear weapons. Code named "Trojan," the plan's top priority was elimination of Soviet refineries, especially those producing aviation fuel, in order to eliminate the Soviet Union's ability to fuel its armed forces.

During the Korean War, the United States made an early decision not to bomb large hydroelectric dams along the Yalu River but reversed the decision two years later in 1952. As Bennett Ramberg, a research fellow at the Center for International and Strategic Affairs at the University of California at Los Angeles, has pointed out in a study of nuclear power facilities and war, "the decision was reversed . . . when negotiations deadlocked and destruction of the plants seemed necessary to hasten the war's conclusion and to make more difficult the repair work the Communists were doing in small industrial establishments and railway tunnels." On the other hand, in Vietnam electrical and other energy facilities were never a major strategic commitment. As with Japan, Vietnamese power plants were too small and scattered to be primary targets. In Vietnam, as in Japan, decentralization preserved substantial capacity.

Israel's 1981 bombing of an Iraqi reactor shocked the world and introduced the energy war to the strife-ridden

Middle East. However, the June bombing of Iraq's Tuwaitha nuclear research center, ten miles from Baghdad, came as no great surprise to astute observers. Tuwaitha housed a new 70,000 kilowatt French "Osirak" research reactor. In 1979, a saboteur's bomb damaged the $275 million reactor, which was awaiting shipment in France. A senior scientist in the Iraqi program was murdered in 1980 in Paris. France blamed Israel, knowing that country's fear that Iraq would use the reactor to build Arab nuclear bombs.

On September 30, 1980, following the delivery of nuclear fuel to the Tuwaitha center, Iran bombed the facility, further inciting speculation that the raid was engineered by Israel, not Iran. But no proof was offered.

The 1981 Israeli attack on the reactor, code named "Operation Babylon," was a precision air raid using U.S.-supplied F-16 fighter bombers. A French technician who observed the raid said that "the precision of the bombing was stupefying." The Tuwaitha center was quickly reduced to rubble.

The attack underscores not only the temptation to bomb centralized power plants, but also the utter failure of international measures to reduce the spread of bomb-grade materials. Israel emerged as the only nuclear power in the Middle East, having already developed her own weapons capability. Ironically, Iraq is a signatory to the Nuclear Non-Proliferation Treaty—which Israel has steadfastly refused to sign— and a United Nations team had recently inspected the Tuwaitha center.

TERRORISM

Expatriates from Libya recently reported that Colonel Qaddafi has employed the former head of East Germany's secret police to train six thousand Libyans in the arts of terrorism and sabotage. Accurate or not, such reports highlight a growing menace in the world. Terrorism is on the rise around the globe; the United States has so far escaped this scourge to a great degree. Terrorism cannot be compared with conventional crime—it is an act directed against all of a society, deliberately designed to shock, dismay, and enrage. Sabotage, on the other hand, while usually political in motive too, is functional in its goal—to destroy a particular capacity.

There is little doubt that the overhead transmission lines and the pipelines that carry electricity and fuel in this country are vulnerable to terrorist attack, particularly the switching centers and aboveground valves. They are essentially unguarded, vulnerable to a variety of weapons, and difficult to repair. Less vulnerable are refineries and power plants. They are guarded—often by personnel trained in counterterrorist tactics—and are generally designed to withstand accidental, if not intentional, damage. Even so, they remain viable targets for internal attack.

Terrorists have attacked energy facilities in the past. In 1972, Black September groups blew up two gas processing plants in Rotterdam; these were the first energy-related attacks by the seventeen largest terrorist organizations that operated between 1968 and 1978. There have been other groups and many targets, as table 2, from the U.S. Department of Energy, demonstrates.

THE BEAR MOVES SOUTH

In the realm of science there is a procedural rule called the Law of Occam's Razor. It says, in brief, that if there are several hypotheses that fit all the facts, you should take the simplest hypothesis and use it until some other facts crop up to show that a new hypothesis is required.

There are a number of facts that call for a hypothesis about today's Middle East.

• A year before the Russian Revolution, Lenin published a tract called *Imperialism,* which enunciated the principle that Western capitalism could not survive without sources of cheap energy and raw materials.

Table 2. Sabotage to Domestic and Non-Domestic
Energy-Related Targets
1970 to Mid-1980

Target	Incidents	
	Domestic	Non-Domestic
Powerline	55	48
Power station/substation	43	21
Pipeline	27	54
Petroleum/gas storage	15	15
Nuclear energy peripherals and support	15	32
Refinery	6	12
Oil well	5	1
Hydroelectric facility	4	2
Mine	2	2
Coal train	1	—
Nuclear weapon association	1	—
Oil tanker	—	3
Oil trains	—	1
Nuclear waste freighter	—	1
Totals	174	192

• In a meeting of the Warsaw Pact nations in Prague in 1973, Leonid Brezhnev stated that the Soviet goal was to dominate the world by 1985 and that the route to this dominance was to control Europe's source of energy and raw materials, reducing that continent to the role of a Soviet hostage.

• In 1979, the Russians invaded Afghanistan for the stated reason that the unrest in the Moslem nation might generate similar unrest among the (very much discriminated against) Moslem population in Russia.

• The Soviet Union has never had the slightest difficulty in putting down restless populations within its borders.

• The Russians' presence in South Yemen is almost total. Yemeni and Palestinian workers are the dominant work force in Saudi Arabia.

• For two decades the Soviets have been bringing students to the USSR from Baluchistan, an area in Pakistan athwart the Iranian border. The southern border of this province is the Straits of Hormuz in the Persian Gulf, the narrow tunnel where a vast amount of Middle East oil makes its way to Japan and the West.

• It is a cliché of Russian history that that country—czarist or communist—has always hungered for a warm-water port.

These facts, quite aside from the well-known build-up in recent years of Russian military capability on land and sea, cry out for a hypothesis. Norman Podhoretz, in *The Present Danger*, argues that, "After more than sixty years of Communist rule, the only thing the Soviets seem to be good at is producing nuclear bombs and missiles. The Soviet Union is still unable to feed itself despite vast expanses of fertile land, and it still has to apply to the United States for technological help in developing other industries. . . . In short, the reason Soviet imperialism is a threat to us is not merely that the Soviet Union is a superpower bent on aggrandizing itself, but that it is a Communist state, armed, as Sakharov says, to the teeth, and dedicated to the destruction of the free institutions which are our heritage and the political culture which is our glory."

There is but one single, prudent hypothesis. We are about to see one of the greatest international conflicts of our time —the energy war. The United States and the Soviet Union are already engaged, and the opening signals affect both societies: increasing demands but decreasing domestic supplies of key sources, such as petroleum. Both societies, as a result of major impending shortages, are vulnerable. What we show in the remaining pages of this book is the inability of conventional policies to protect the nation from becoming vulnerable—both to energy shortages and to war. What we shall offer are some ways to reduce U.S. vulnerability, while, at the same time, preventing the energy war from enveloping our planet in a nuclear holocaust.

3. Two Hundred Twenty-four Million Hostages

THE PRACTICE of civil defense used to be a relatively straightforward exercise. Before the bombing raids, the local civil defense sirens would shrill a timely warning and the nearby populace would take cover in underground bomb shelters. Property would be damaged, if not lost, but people would emerge from the earth and rebuild.

Today, however, the idea of civil defense is a controversial subject—not just controversial in the sense of its adequacy, but controversial on the grounds of its *applicability* at all in the nuclear age. Since the Second World War, most of the industrialized nations that would be targeted by nuclear weapons have put into effect some form of civil defense. Leaders in the fallout-shelter race are the Russians and the Swiss, who have built elaborate, well-financed underground redoubts with emergency food, water, energy sources, and medical and communications supplies in the event of World War III.

AMERICAN-STYLE

The approach to civil defense taken by the United States in the nuclear age began to develop soon after the Second World War. At that time U.S. government studies based on the Hiroshima and Nagasaki explosions indicated that shelters could save a substantial part of the population from the blast effects of the bomb. In 1950, the Civil Defense Act was passed, and a large national effort was mounted to plan for installation of the shelters. However, when the fifteen-megaton "Bravo" test was conducted by the United States at Bikini atoll, the planners learned that radiation was a bigger problem than they had initially anticipated. The thermonuclear weapon tested at Bikini created a fallout-laden, 100,000-foot mushroom cloud that was carried by wind to nearby inhabited atolls and to the Japanese fishing vessel *Fukuryu Maru (Lucky Dragon)*. The Atomic Energy Commission (AEC)—in charge of the nuclear test—immediately evacuated the atoll-dwellers, who were not allowed to return to the island for several years.* The twenty-three Japanese fishermen, who were one hundred miles downwind of the blast, suffered excessive radioactive contamination (one died), and the Japanese government seized tuna catches from over six hundred boats, causing a general panic in that island nation.

These revelations caused civil defense planners to rethink the blast shelter concept, and population relocation to "safety zones" became the trendy civil defense notion of the 1950s. Today, this remains the only real civil defense program left in the United States, and even it is a plan on paper only. Sidney Lens, in his provocative history of the nuclear arms race, *The Day Before Doomsday,* notes that the "evacuation panacea" of the late 1950s created the new national interest in civil defense, capitalized on by the Kennedy administration in the early sixties. "Evacuation arrows dotted the highways from coast to coast, offering the illusion that in the ten or twelve hours it took the Soviet bombers to get here there

---

*In its inimitable fashion, the AEC issued a false press release on the "Bravo" test, stating that the evacuation had been taken "according to plan as a precautionary measure." There was, in fact, no plan; the islanders were exposed to substantial radiation, causing lesions and scars.

would be plenty of time to head for safety a few dozen miles from Los Angeles, Chicago, and New York. As late as January 1959 New York State's civil defense director, who had evidently never seen a morning expressway jam, was still claiming that 'every person in New York City could be moved to the country within 50 hours.' "

The philosophical underpinning of the revived civil defense effort was written by Herman Kahn in 1960. In his classic work *On Thermonuclear War* he claims that with a well-financed civil defense program "we could evacuate our cities and place our forces on a super-alert status, and thus put ourselves in a much better position to strike first and accept the retaliatory blow. We might then present the Soviets with an ultimatum."

Along with the renaissance of civil defense projects during the Kennedy era came an increase in the annual budget for the civil defense agency: it soared from $60 million to $700 million.

Suddenly fallout shelters were again being built; a new trade association was formed by the forty-odd manufacturers whose presidents predicted that the fallout-shelter market represented a potential $200 billion in sales. And yet public leaders were concerned that citizens might not have time to make it to the shelters in the event of a missile strike. Missiles launched by submarines can reach U.S. targets in less than ten minutes. In the resultant chaos the populace would be uncontrollable. California Governor Edmund G. "Pat" Brown said in an interview at the time: "Recently, the public has been alarmed by groups and individuals who say they will resort to violence in the event of nuclear war. These guerrilla bands have announced that they will not hesitate to shoot down citizens entering areas designated by the selfish as their own. . . . This reversion to cave-man barbarism is a cause of concern to every thinking American." In other words, if you got to your fallout shelter or your mountain hideaway a bit too late, you might encounter an unfriendly band of rifle-toting strangers.

Even while the Kennedy administration was expanding the civil defense program (and the defense budget generally), neither the president nor his key advisers showed much enthusiasm for perfecting the quality of civil defense programs insofar as they could protect the population from an enemy attack. President Kennedy noted in one speech that the United States would deter an enemy attack "only if our retaliatory power is so strong and so invulnerable that [the enemy] knows he would be destroyed by our response. If we have that strength, civil defense is not needed to deter the attack. If we should ever lack it, civil defense would not be an adequate substitute."

As a result of the Kennedy nuclear policies, and the clear concern of the administration that civil defense offered little as a post-attack palliative, only the first two years of the program were funded to provide federal support for fallout shelters, emergency supplies, and communications equipment. This trend has continued to the present. The national civil defense budget has limped along with appropriations of around $100 million per year since the heyday of activity in the sixties. Most of the money is spent to administer the civil defense bureaucracy and to write up fanciful paper plans for the "crisis relocation" of urban masses to the countryside, which will be discussed below. The last director of the Defense Civil Preparedness Agency, Bardyl R. Tirana, was outspoken in his contempt for the government's attitude toward civil defense and lectured congressional committees on the absurdity of granting the agency a budget simply to keep the sign posted on the door. "It's folly," he said, "to think of civil defense as being able to protect society as we know it from destruction in a very large-scale nuclear exchange. But subject to policy approval, "civil defense could make a minor or small contribution to stability and deterrence."

In 1979, the Defense Civil Preparedness Agency was disbanded and became one of the disparate elements of the federal government fused into the new Federal Emergency Management Agency. After nearly two decades, the initial programs had fallen into such disuse that one had trouble locating even the studies in the FEMA library—such was the state of affairs that money had not even been appropriated to provide an adequate library for the agency.

CAN CIVIL DEFENSE WORK?

The most vexing question about civil defense is whether it will work, and whether the program is worth funding at any level. The goals of civil defense were recently summarized by the Congressional Office of Technology Assessment as follows: "Civil defense seeks to protect the population, protect industry, and improve the quality of post-attack life, institutions, and values." Thus, the program has several vital goals—not only to mitigate the effects of a nuclear attack, but also to provide for some form of physical and institutional continuity for the nation's survivors.

A major emphasis of civil defense ever since the first U.S. program was outlined to Congress by Defense Secretary McNamara in August, 1961, has been the identification (and, during the 1960s, construction) of fallout and blast shelter spaces. Even though the Kennedy plan was ambitious and involved sending experts to shelter spaces in order to map them and to stock them with food and key materials, the program had shortcomings. As McNamara told the Congressional Joint Atomic Energy Committee, "This does not mean the program will save 50 million lives. Again . . . nearly 75 percent of the deaths from the hypothetical attack would have resulted from blast and thermal effects." The national program of fallout shelters would not alleviate this, he said, but they would "save at least ten to fifteen million lives."

A classified study of the costs of projected U.S. civil defense programs was performed in the late 1970s for the Department of Defense, and an unclassified summary was circulated by the Defense Civil Preparedness Agency. A major finding at the time was that the civil defense program of crisis relocation could save only 30 percent of the population from the initial effects of a large nuclear attack on the United States. The study noted that a better-funded program of crisis relocation would save 40 percent of the population and would cost $1.6 billion per year—about ten times what was spent on federal civil defense in the late seventies. The DCPA felt their suggested program was the ideal civil defense effort because ". . . based on extensive research and developmental work . . . crisis relocation could be highly effective—given the requisite planning and development of supporting systems and capabilities, and given about a week for moving and protecting the bulk of our population at risk." The other alternative cited in the classified analysis of civil defense was a $60-to-100 billion program to provide blast shelters in major cities; although the agency believed that this would assure 90 percent population survival (from initial effects), the program was ruled out as too expensive.

Citing the successful evacuation of 1.5 million women and children plus 2 million additional voluntary evacuees from London in three days in 1939, the DCPA concluded that a well-funded relocation program would be the best possible plan of action for the nation. This notion is theoretical at best and assumes that nuclear warfare will conform to a seventeenth-century model of genteel sparring—in which case both sides would be aware of the impending holocuast and would spend a week or so apiece relocating massive numbers of people to remote areas or shelters for protection. In an age in which missile attack from submarines can occur with startling devastation in only a few minutes (or in half an hour from ICBMs), the one-week assumption postulates a hardly believable period of pre-attack calm, during which the president would act on relocation plans.

However, the perils of poorly planned and inadequately funded relocation plans were pointed out in a 1980 report to the Federal Emergency Management Agency (FEMA) by one of the nation's best-known civil defense analysts, William

M. Brown of Herman Kahn's Hudson Institute. The Brown study, entitled *Postattack Recovery Strategies*, surveyed the entire thirty-five year history of federal civil defense efforts and concluded that the current "centralized federal approach" to crisis relocation is fraught with basic problems. "We find it hard," the study stated, "to place any great confidence in the outcome of a conceptual model of planned responses under federal control; if such circumstances are followed by a nuclear attack we have a rather chaotic and gruesome image of what is likely to happen if the civil defense preparations at the start of the (prenuclear attack) crisis are as meager as they have been heretofore. An analogy which may clarify this image is one in which a yet unrecruited army is to be used for a major battlefield task. It is to be recruited in a few weeks and sent into battle in an unprecedented situation with little or no equipment, training, supplies, or experienced officers. Under such circumstances, could we expect the hypothetical army to maintain order, let alone confront the enemy and emerge with a victory?" The U.S. civil defense structure today, the study notes, "is still without even a rudimentary program for dispersing the population, for providing sufficient survival supplies, for creating an adequate amount of suitable shelter against either blast or fallout threats, or for rescuing, reorganizing and recovering once the hypothetical attack has ended."

In real terms, the existing crisis relocation programs for urban dwellers seem fanciful beyond belief. We found in our research that the federal relocation planners consider California's war preparation efforts and those for Washington, D.C., the best in the nation. The "California War Emergency Plan," first published by Governor Ronald Reagan in 1970, is currently being updated to cover a comprehensive crisis relocation element. This plan provides for an elaborate division of state responsibilities, with the National Guard taking charge of welfare, law enforcement, and general assistance to cities in conjunction with the California Highway Patrol and other branches of state and local government. The plan delineates specific time periods for preparation, including the "Preparedness" period and the "Attack" and "Early Post-Attack" periods. On analysis, however, we found that the plan to relocate the Riverside area of Los Angeles, considered (again) one of the most complete in the state, had no provisions for food, fuel, or transportation. The Riverside evacuees, who are supposed to relocate in Indio, a small town in the San Bernardino mountains east of Los Angeles, are expected to provide their own food, emergency supplies, and transportation. The plan does not specify who will supply fuel in the advent of a rapid evacuation, whether or not there will be adequate space or supplies in Indio, or whose exact jurisdiction the evacuating hordes will be turning to—the state government, the local government, or the citizens of the town of Indio, who are, for the most part, not even aware that they are a "host city." Even the radio frequencies in key police and emergency vehicles are not compatible; the radios, therefore, in state police and highway patrol cars cannot receive transmissions from the local police, who supposedly would be on the road directing the biggest traffic movement in California's history. Even carrier pigeons have been suggested. The problem of incompatible communications frequencies is compounded by the fact that EMP effects of nuclear weapons will render most of the radio systems inoperative in the first microsecond of the nuclear attack. An illustration in the Defense Department's "Attack Environment Manual" actually suggests, albeit humorously, using birds for communication "until the EMP threat is over."

The Washington, D.C., plan is equally ridden with problems. A recent *Washington Post* survey of the situation by Ann Cooper points out that some 80,000 evacuees from the nation's Capitol will be sent almost two hundred miles to a sleepy Virginia mountain resort town, Hot Springs, which has one 694-room hotel, a golf course, and a lot of "open area." As the author points out, " 'open area' is not very good to be in in a nuclear attack, since communities that are not enemy targets still stand a good chance of having radioactive fallout blown in from elsewhere." Some 275,000 carless Washington residents are supposed to be bused to rural areas away from the Capitol, but local civil defense officials have no idea who will be around to drive the buses and are counting on volunteers. On a more specific and personal level, the Washington civil defense office offers a list of articles to be taken by fleeing residents, including all the food you can carry, water, clothes, bedding, soap, sanitary napkins, first-aid kit, portable toilet, prescription drugs, hammer, saw, screwdriver, nails, stocks, bonds, cash, and your social security card. A second civil defense list prohibits guns, alcohol, and drugs. One wonders which one will be heeded.

The Brown study effectively summarizes the Achilles heel (at least, one of them) of the current federal crisis relocation program: "Following large attacks, food production and distribution could stop for several months, or in a massive, highly radioactive attack perhaps for a year or more in most of the country, and then might only slowly recover. . . . The country is currently fortunate in having tremendous stocks of food, perhaps enough for 1½ or 2 years; but unfortunately they might not be easily available to the population, post-attack." The study offers the example of Orleans county, New York, which could provide food for local residents for forty-two days, using existing stocks. However, when evacuees are added to the local population, the existing food stocks would last less than a week. Among the study's many recommendations for revitalizing civil defense is that the federal government should immediately take actions to store critical materials, such as food, water, fuel, and medicine in the relocation areas.

Even so, the Brown study questions whether government —as we know it—would survive at all. In a remarkably understated phrase, Brown notes that "there are reasons to think that the federal government actually might be quite vulnerable," citing the centralized nature of modern U.S. bureaucracy and its concentration in Washington, D.C., and other key locations. To remedy the problem of a vaporized democratic civilization, the Brown study suggests that the nation may need to create a federally funded, independent "professional survival-oriented cadre distributed among the nation's localities." This one thousand-to-ten thousand-person paragovernmental force would exist for the sole purpose of helping communities cope with a nuclear war crisis; in effect, the cadre would be a paragovernmental priesthood— the technical arm of government in the event of war. Indeed, these professionals might even become the government. The image conjures up the suggestion made by physicist Alvin Weinberg, who would create a sort of "nuclear priesthood" whose job it would be to take care of the nuclear power establishment and its radioactive wastes for the countless human generations required to manage the residues until the radionuclides lost their potency and poison, which it is estimated would last thousands of years.

SWISS PRECISION

Although the United States is mainly relying on population relocation for civil defense, both Switzerland and the Soviet Union have opted for elaborate underground blast-and-fall-out shelters. The Swiss have developed blast shelters that can protect six million people, about 90 percent of the population, from the initial effects of a European nuclear war. The Swiss government views population evacuation plans with some contempt, and the official civil defense manual states the case as follows: "Transportation of the people into the receiving areas and adequate supplies could not be guaranteed under war operations. Furthermore, such evacuation activities could hinder important general defense actions. The uncertainty regarding time and duration of such evacuations would render the operation especially difficult. Consequently, large-scale transfers of people in a modern war in this country are ineffective and even dangerous and must be avoided. This is feasible on condition that each inhabitant is provided with a shelter place at or near his domicile."

Peter Laurie, author of an authoritative analysis of civil defense, *Beneath the City Streets*, notes that Switzerland spends one-quarter of its defense budget on civil defense programs, following a national referendum on the subject in 1962. In addition to commenting on the nation's elaborate system of underground bunkers, he observes that "there is a parallel organization for industry which protects vital processes and staff; systems for the railways, telephones, broadcasting, and a huge program for microfilming documents, ranging from electoral rolls to ancient manuscripts. There is a department which photographs old buildings in minute detail, so that the historic monuments of Switzerland shall rise as good as gold from the ashes." After a tour of this "second Switzerland," he said that "The mind boggles at the scale, the thoroughness, the standardization. It is evidence of a paranoia of a very high and determined order—using the word in no insulting sense, but rather to mean the art of ignoring the reassurances of those who do not necessarily mean you well. . . . There is no doubt that if there is a war in Europe, we shall lose most of our resources and 40 percent of our people, while the Swiss will be almost unscathed in everything that matters. Assuming the major powers cripple each other—as they have the means and the will to do—postwar Europe will be run by the Swiss and the Swedes, the only countries with respectable civil defense. The mind boggles at that, too."

SOVIET-STYLE

Although Russia does not have as comprehensive a civil defense program as the Swiss, this nation has devoted considerable effort to establishing a strong program combining the

use of relocation plans and well-equipped blast shelters. Unlike the United States, the Soviet civil defense program is
under military control (under the leadership of a deputy
defense minister, Colonel-General A. Altunin) and has over
one hundred thousand full-time employees at almost all levels of government. (The United States employs a few hundred people in this enterprise.) According to a 1978 Central
Intelligence Agency assessment of Soviet civil defense, the
Russian program would cost about $2 billion yearly, if du-
plicated in the United States. About three-fourths of this
would be to pay for the manpower necessary to implement
the program.

The CIA believes that the effectiveness of the Soviet program in reducing casualties in an attack would "depend primarily on the time available to make final preparations before an attack." Under the worst attack conditions (a major
counterforce attack), total casualties would be over 100 million, but a large percentage of Russian military, government,
and economic leaders would survive, due to the existence of
convenient and well-built blast shelters designated for about
110,000 "key" people. With two to three days for evacuation,
the Soviets could reduce total casualties by 50 percent, and
"under the most favorable conditions for the USSR, including
a week or more to complete urban evacuation and then to
protect the evacuated population, Soviet civil defenses could
reduce casualties to the low tens of millions," says the CIA.
However, the study notes that "massive damage to their
economy and the destruction of many of their most valued
material accomplishments" could not be avoided in a major
nuclear war.

The serious Soviet attitude about civil defense has
prompted a number of proposals for a U.S. duplication of this
effort. A leading advocate of increased civil defense programs, T. K. Jones of the Boeing Corporation, has catalogued
the various population and industrial protection measures
implemented in the Soviet civil defense effort. For example,

he notes that the Soviets have implemented shelters for protecting the critical elements of their industrial work force, as
well as protecting critical industrial machinery from nuclear
blast effects. In a 1976 congressional hearing, he stated,
"These Soviet preparations substantially undermine the concept of deterrence that forms the cornerstone of U.S. security. . . . These defensive preparations, combined with the
increasing power of Soviet strategic offensive forces, have in
fact destabilized the strategic relationship between the two
nations." Concurring in this opinion is Leon Goure, a campaign adviser to President Reagan, who says that "the Soviet
civil defense system would save the lives of all but 10 percent
of their population in a nuclear war. This means that in effect,
we have entrusted our survival to the good intentions of the
Politburo of the Soviet Union." Noting that it would cost the
equivalent of $350 per capita for the United States to emu-
late the Soviet program, he asks, "Is an American worth
three hundred fifty dollars?"

There is a lively debate, however, over the effectiveness
of the Soviet civil defense. According to a 1978 survey by the
U.S. Arms Control and Disarmament Agency (ACDA), the
Soviet Union could actually do little to mitigate the effects of
a major nuclear attack. ACDA asserted that even though
evacuation plans might reduce the fatalities to the 25–35
million range, the evacuees themselves could be targeted by
ground-burst nuclear weapons, killing 70–80 million people.
ACDA also challenged statements by Boeing and the U.S.
Defense Nuclear Agency concerning the effectiveness of
hardening industrial equipment, stating that "attempts to
harden aboveground facilities are a futile exercise, and that
even buried facilities which are targeted cannot survive."

Fred M. Kaplan, a researcher at the MIT Center for International studies, also discounts the effectiveness of Soviet
civil defense, pointing out that even the official Soviet civil
defense manual, *Grazhdanskaya Oborona,* consists mostly of
theoretical plans and goals. (The section of this manual on

nuclear weapons effects is a poorly translated chapter from the U.S. official manual on nuclear weapons.) Kaplan points out that the much-touted Soviet civil defense goal of industrial dispersion, called "a decisive measure for insuring the viability of the economy in wartime" by Colonel-General Altunin, is self-contradictory, given the fact that the USSR has an "overwhelmingly concentrated" industrial economy developed methodically since World War II. Kaplan notes that starvation would be the major Soviet problem following the war: "Fallout would seriously affect farmland; in any event, Soviet agriculture, highly dependent on chemicals, could not be readily regenerated if any of the 25 chemical factories in the Soviet Union are blasted."

POST-ATTACK: WHAT'S LEFT?

The most vexing, and as yet unanswered, question in the whole civil defense debate, is that of recovery after a nuclear attack. Assuming that we would have a reasonable period of time (one to two weeks) to evacuate the targeted areas before a nuclear war broke out and assuming a big federally financed program for stockpiling critical supplies of food, fuel, medicine, water, and other essentials in "host" areas for migrating hordes, it is conceivable that hastily erected shelters could protect some of the evacuees. Such shelters have been proposed by civil defense officials, and shelters that are capable of being built in forty-eight hours could protect "tens of millions of Americans following field-tested, written instructions," argues Cresson H. Kearny in his well-researched handbook, *Nuclear War Survival Skills,* published by the Oak Ridge National Laboratory.* Such shelters would offer protection from fallout but not from a nuclear blast; their use

*The Kearny handbook is available from the National Technical Information Service, 5285 Port Royal Road, Springfield, Virginia, 22161 for $10.75. It is the only complete manual of its kind available in the United States.

would be limited to rural areas and cities not targeted by large nuclear weapons.

The uncertainties surrounding the survival of the population in a nuclear war warrant attention to the whole notion of civil defense. The study by the Office of Technology Assessment points this out by concluding that "no effort has been made to answer key questions [such as] 'Would a civil defense program on a large scale make a big difference, or only a marginal difference, in the impact of a nuclear war on civil society?' " Most of the research we have evaluated here indicates a high level of government optimism that everything will work out according to plan: early notice of the attack, well-executed evacuation, expeditious shelter-building, no violence in the "host areas," good weather, and togetherness.

Of course, what is likely to happen is a failure in some and perhaps all of these systems. Even the effects of EMP from one or two well-placed atmospheric nuclear blasts would wreak havoc on preparations for population evacuation, due to the simple fact that almost all new civilian autos and trucks are equipped with "black boxes" controlling electronic ignition systems and controls. Combine this with a car antenna, which would conveniently collect EMP, and one can foresee the simultaneous failure of most of the national vehicle transportation systems. That, of course, does not count airplanes, computers, radio stations, telephone lines, and the control systems of oil, coal, and nuclear power plants. In this eventuality, burning out a few million automobiles would be somewhat less disastrous than knocking out the controls of the nation's seventy-odd nuclear reactors.

Roger Rapoport, in *The Great American Bomb Machine* argues that the scores of civil defense studies on plagues, energy, food, transportation, government, and other aspects of recovery serve two functions: "First, they keep the government happy by advancing the fiction that nuclear war is survivable, so generals and politicians can play with their 'nuclear option.' Second, they keep the postattack researcher employed. Were he to use valid scenarios of all-out war, instead of token nuclear attacks, there would be virtually nothing left to evaluate. But, by claiming that nuclear war is survivable, he validates the need for his own work and opens the way to further research that keeps him on the public tit."

The truth is probably that a nuclear war, even an "all-out" nuclear war, would not totally destroy civilization, but just what would be left is open to question. Given the poorly supported optimism of most official studies of civil defense, especially the newer studies that attempt to paint an orderly picture of crisis relocation, it is exceptionally difficult to gauge the logic of such a national effort. Twenty years ago, the chief of environmental sciences for Atomic Energy Commission, John N. Wolfe, told a congressional committee that the effects of nuclear war "are awesome to contemplate." Wolfe noted that the combined effects of multiple nuclear blasts, ensuing fires, and deadly radioactivity would "create vast areas that would be useless to the survival of man. . . . Fallout shelters in many areas seem only a means of delaying death. . . . With an environment so completely modified, the question is, where does man go after his sojourn in shelters? What does he do upon emergence?"

Wolfe told the committee that ecological devastation would especially affect key plants and animals necessary for human support systems. Trees such as pine, spruce, and fir are much more susceptible to radiation effects than are many broadleaf varieties. In such forested areas, the destruction of the forests might trigger fires and floods, rendering the environment a wasteland. Many animal, microorganism, and insect vectors of diseases, such as bubonic plague, dysentery, malaria, typhus, and so forth, are less susceptible to radiation than humans. "The point is," he said, "that nuclear war triggers a whole series of effects. If you killed bumblebees you'd have a hard time growing clover. You wouldn't get pollination. If you destroyed the bird population you'd get an explosion of insects that birds now keep under control. I can multiply these examples a thousand times. . . . Maybe not one of these things will be difficult to handle by itself, but when they all come at once and when there are problems of food, clothing, and shelter as well . . ." He did not complete the thought.

One is inclined to reluctantly conclude from the overall survey of civil defense that a massively financed (100 billion range) U.S. program may preserve one-third to one-half of the American population from the immediate effects of nuclear weaponry, but the world above ground would not necessarily be capable of supporting much in the way of civilization following the holocaust. After all, we cannot, for any amount of money, create a "Noah's Ark" of the diverse ecosystem in underground bunkers.

In fact, the best system of civil defense is, in our view, one that is designed primarily to prevent a nuclear war by reducing the activities in the national sphere that would create conditions tending toward war and by, as an ancillary benefit, decentralizing the nation's key activities to such an extent that nuclear targeting of central points would become a meaningless pursuit. As John Jay wrote in the *Federalist Papers*, "the safety of the people of America against dangers from *foreign* force depends not only on their forebearing to give *just* cause of war to other nations, but also on their placing and continuing themselves in such a situation as not to *invite* hostility or insult; for it needs to be observed that there are *pretended* as well as just causes of war."

In the name of "mutually assured destruction" (MAD), the civil defense theory that finally emerged in the sixties, our government has given up the people of the United States for lost in the nuclear age. The government has deliberately planned the almost total sacrifice of the civilian population in the name of nuclear deterrence. As Nigel Calder observes in his forthright book on the current dangers of nuclear war, *Nuclear Nightmares*, "it is not playing fair, in the game of 'mutually assured destruction,' to try to minimize the effects of the other fellow's strike. Indeed, if you start building strong shelters and drawing up plans for evacuating your cities, you could upset the balance of terror. Your actions imply, at best, that you do not have unlimited faith in deterrence; at worst it means that you are trying to avoid 'assured destruction' on your side in order to be able to fight and win a nuclear war."

Redefining civil defense should be a national objective, given the interest of the current president (and of previous presidents as well) in using civil defense as a deterrent and survival option in the case of "limited" nuclear war. What is needed is nothing less than a sober view of the reality of defense in the nuclear age, requiring an approach based on actions that could mitigate, if not preclude, a nuclear war.

4. The Vulnerability of America

WE HAVE a highly centralized energy system in the United States. The electricity you use in your home derives from an increasingly centralized system of large generating plants. These in turn depend on increasingly centralized systems of fuel production, transportation, refining, and storage. Even the people who bring fuel oil to your house, with whatever regularity you can arrange with them, are likely to be large entrepreneurs—not local acquaintances.

What do we mean by *centralized* as opposed to *decentralized?* Such terms are relative, as are the terms *large-scale* and *small-scale,* but it is possible to think of centralized power as the dependency of an energy system on a relatively small number of large components. A nuclear power plant is a very large piece of work, something that the architect of the Taj Mahal would approve of: an awesome structure. As electric power plants and their service areas increase in size, a centralized dependency emerges. And so does vulnerability.

The U.S. energy system is itself extremely energy-intensive. In every stage of operation, today's energy system absorbs a great deal of energy just to sustain itself. One-fourth of every barrel of oil produced is lost in the process of refining it.

Throughout the physical world, such losses are inevitable, since processes of energy conversion are inevitably accompanied by *entropy*. Energy constantly moves from a higher to a lower state; in the operation of an engine, energy is not destroyed, but the waste heat resulting from the operation of the device is a lower quality form of energy than the original fuel. The science of thermodynamics teaches us that this process of energy degradation (or, entropy) is unavoidable.

Entropy affects all natural systems—not just the operation of heat engines. Economist Nicholas Georgescu-Roegen notes that, in "entropy terms, the cost of any biological or economic enterprise is always greater than the product." The energy contained in a gallon of oil can be used for a number of purposes, ranging from the manufacture of petrochemical products (fertilizers, drugs, clothing, and so forth) to direct combustion to produce electricity. The state of the original oil is referred to by scientists as "low entropy," but when the coal is converted to chemicals or electricity, the residual energy cannot be re-used to make more products or power. In this dilute form, the energy is in a "high entropy" state.

When electricity is made from oil, the oil is burned in a boiler to create heat that raises the temperature of water, making steam that rotates a turbine to generate the electrical current. This process results in the production of electricity, a quite useful form of energy, but it also creates the loss of two-thirds of the original stored energy of the oil.

The United States uses about 7½ million barrels of oil every day to produce electricity. Five million barrels are lost irretrievably to the exigencies of entropy. If you add all that up, it amounts to 60 percent of all the oil imported into this country that goes off into the atmosphere—or the water cooling systems—as useless heat. You cannot avoid entropy in an energy system, but there are a few measures that can be taken to decrease the loss they will be discussed in the second part of this book. For now it is sufficient to know that we are wasting a prodigious amount of energy and that we have a highly centralized energy system, which not only makes dealing with entropy all the harder, but also provides an enemy with fewer but far more important targets.

How did this come about?

THE URGE TO CENTRALIZE

A hundred years ago, aside from animal and human brawn, wood was the primary source of energy in America. Up to 90 percent of the nation's energy requirements were supplied by trees until coal began to replace the overcut forests as a source of energy. (Coal, like the wheels on Aztec toys, was recognized for its usefulness long before humanity at large decided it was a necessity. The Hopi Indians of northern Arizona used coal in 1000 A.D. By the time the Spaniards reached the area, nearly 100,000 tons had been mined. It was used for firing pottery, but because it smelled so bad the Hopi kachina gods apparently banned it from homes.)

Wood was an easy source of energy for the Europeans arriving in America, and until they had decimated the forests by the mid-nineteenth century, there was no need to use coal. Then, the iron and steel industries began in earnest, and coal became the preferred boiler fuel. By 1885 more coal was used than wood. By the turn of the century coal supplied more than two-thirds of U.S. energy needs and remained the dominant fuel well into the twentieth century. By the time, however, that the South seceded from the Union, a discovery had been made that was to fundamentally change civilization. In 1859, a man named Edwin Drake drilled America's first oil well in Pennsylvania. The product was called kerosene, to distinguish it from the coal-based fuel then in use. Within a year, American oil wells produced 500,000 gallons of kerosene; in ten years they were producing four times that.

Even more dramatic changes were yet to come. At the turn of the century, there were 8000 automobiles in the United States. Eight years later there were 194,000. By 1930,

23 million vehicles, using gasoline, were registered in America. In the homes of these increasingly mobile Americans was a growing array of appliances—made possible by the ever-increasing availability of electricity. Larger power stations, with higher efficiencies were built, and the lower costs to consumers further stimulated the use of electricity. Until recently, starting with World War I, the use of electricity doubled every ten years. Since 1945, electricity use has sextupled; it now provides a quarter of American energy. The use of natural gas has increased astronomically. World War II, it is said, was an oil-based war, and America was the chief supplier of oil to the Allied war effort. For every two tankers sunk by the enemy—and some six hundred were sunk—the oil rich American economy produced four tankers. Oil wells were drilled, and production facilities, refineries, and synthetic rubber plants were built. Oil—along with strategy, courage, and human imagination—won that war. And while oil was winning the war, a team led by Enrico Fermi in Chicago built the the world's first nuclear reactor.

After the war, U.S. energy use expanded dramatically. Energy-rich mechanical heating and cooling technologies became standard equipment for homes and commercial buildings. The greatest public works project in the history of mankind—the Interstate Highway System—was built, extending the range, and the fuel consumption, of the automobile. Railroad trains as freight carriers and passenger trains gave way to trucks and airplanes. It takes, by the way, two barrels of oil to make one barrel of jet fighter plane fuel. Between the war's end and 1969, the U.S. population increased 43 percent; the use of motor fuel increased by 100 percent. In short, a very expensive thirty-five years have brought us to our current pattern of energy use and production. Much of this had to do with what economists call economies of scale. From chopping wood, building small hydroelectric dams, and mining coal, more or less locally, to producing dribbles of oil for funny-looking cars, the energy system of the United States became a major, centralized enterprise. Transmission lines snaked across locales and regions, as did pipelines. Larger plants for generating electricity, larger mines, and larger corporations became more efficient. Small units were incorporated into larger systems.

Size versus efficiency is a key consideration in any investment project such as a power plant. The idea of economy of scale in power plants involves the use of the long-run average (or unit) cost (LAC). (The irony found in many acronyms should be a subject for further study.) Average cost is obtained by dividing total costs by the level of output, or activity. Long-run cost is emphasized because in the short run, factors that are uncertain can be fixed relatively accurately. For example, a number of technological advances can be predicted with some certainty. Demand, in the short run, is more or less predictable. But these factors become far less predictable as the time frame lengthens.

Economies of scale occur, and long-run average costs fall (which means that returns on investment increase when a number of factors are at work). First, division and specialization of labor and machinery can be achieved with increase in size. Second, to take advantage of the latest technology, production must occur in large quantities. High-volume purchasing and shipping can result in savings (from larger discounts, and so forth). And finally, maintenance and repair crews must be on hand regardless of size of the power plant. There are other factors, including the cost of financing.

The above formulas are viable up to a point, but finally, it seems, the quest for economies of scale is defeated. For example, too many workers may be assigned to a given unit of equipment, or, on the other hand, one worker may be assigned to operate too much equipment. This leads to *dis*-economies of scale. Figure 5 illustrates a typical long-run cost average curve. At a certain size, diseconomies occur and the nicely descending curve begins to go up at an uncomfortable angle. Based on such considerations, it seems that the opti-

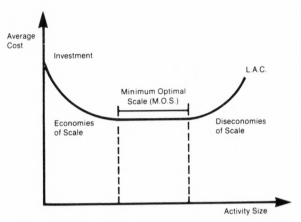

Figure 5. Long-Run Average Cost Curve (LAC)

mum size for a coal-fired steam turbine generating plant is less than 750 megawatts. The optimum size for nuclear plants may also be considerably less than the current 1000–1500 MW models. On the other hand, back in the late 1960s, engineers were predicting that nuclear power stations would reach unit sizes of 10,000 megawatts and that these mega-plants would be of such efficiency that by 1990 most smaller power stations would be obsolete and retired. This has clearly not happened. Despite the fact that these engineering analyses were performed by the most skillful technical people of the time and at considerable expense, they were wrong. Just why this was so will become a bit clearer as we look at individual components of our energy system.

CONVENTIONAL ENERGY

About three-quarters of the American energy supply in 1977 came from a readily available, concentrated, and easily transportable source: oil and natural gas. To find oil is a complicated and expensive process based on trial and error; only drilling can prove its existence. More than one-fourth of the land area of this nation is now under lease for oil and gas exploration. Estimates of how much oil there is left in the ground in the United States vary widely—from 55 billion barrels to 456 billion. At current rates of consumption (six billion barrels a year) the U.S. domestic oil supply will last between one and seven decades. As our own production diminishes, our reliance on imported oil will inevitably grow —even if the much vaunted deregulation provides the oil industry with the incentive to do more drilling. We may be the biggest producer of oil (10.3 million barrels a day), but we are also the world's largest importer, bringing in more than nine million barrels a day. Most of this imported oil comes from the various nations of the Middle East, with a significant component coming from virtually all the other regions of the world: Africa, Indonesia, South America, and the Caribbean. A vast portion of this enters the country at ports centered in a few places, mostly the Gulf Coast. A large fraction of the refining process of this imported crude oil takes place in the Gulf Coast, though virtually all states of the union have some refining capacity.

The petroleum industry is highly centralized. The basic movement of petroleum products, as can be seen in figure 6, is from the Gulf States to the southeastern and north central states by pipeline. It moves from the south to the great industrial centers of the north. All other petroleum transportation pales in contrast to this Gargantuan supply line. The distribution of petroleum products is dependent on a reliable flow of imported and domestic oil.

In the event of a war, such as the strike we described in chapter 1, another series of problems would arise. They can be brought into sharp focus by asking a few questions: Would those foreign nations who own the big tankers that bring oil to our ports wish to incur the wrath, if not the destruction, that could so easily be inflicted on them by the Soviets or any other inimical nation? Would there be any port facilities big enough left to handle the job of off-loading millions of barrels

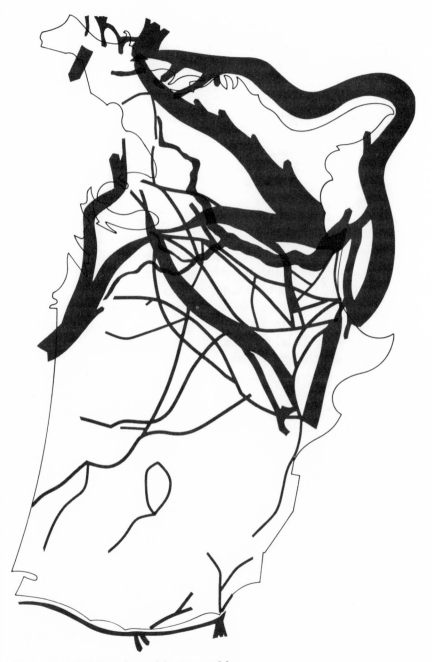

Figure 6. Total Petroleum Movement Map

of oil? Would there be any transportation facilities capable of moving that oil—even if we could somehow refine it—from, say, Louisiana to New York? Add to this the fact that the oil-producing nations (OPEC) know full well that while they own more than half of the known reserves of oil on the planet, they recognize the inevitable: it will run out. As a result, according to a report by energy analyst Fereidun Fesharaki, "OPEC exports are expected to decline from 28.8 million barrels per day in 1979 to 22 million in 1985 and 17 million by *1990.*"

Energy analyst Walter Levy recently commented that "the medium-term prospects for the security and availability of oil and for the economic and political stability and strategic security of the Western world are indeed disturbing. If we now consider the longer-term outlook, the picture looks equally forbidding. Not only are all the economic and political problems to which we have already referred likely to remain with us; but in addition, even by the year 2000, the massive challenge of moving from a mostly oil-based energy economy worldwide to one that can largely draw on other more amply available energy resources will probably still remain largely unresolved." The worsening international situation is the heart of our energy security problem.

There is a double bind here. The oil supply is unpredictable and the way in which we handle both our own and our imported supply is very fragile. Anyone capable of reading this book is also capable of devising a scheme to eliminate some crucial component of the oil-based economy upon which we all depend. A fanciful but terrifying novel, *Shipkiller* told of a vengeful doctor who, equipped with a Swan sailboat and a few rockets purloined in Africa but made in America, destroyed a supertanker. Perhaps out of environmentalist conviction, the author had his hero eliminate this one thousand-yard vessel when it was empty, and it did not cause an embarassing oil slick. Fanciful it was, but it was also a textbook in vulnerability.

Liquefied natural gas (LNG) is brought into this nation in large ocean-going vessels similar to oil tankers—they are basically tanks within tankers that keep natural gas at super-chilled temperatures that concentrate its energy component 600 times, a very efficient way of transporting our second most used fuel—also a very dangerous way of doing so. Such ships, each carrying enough fuel to provide one year's electricity for a city of 85,000 people, are awesome. A one-thousand-foot long LNG tanker heading for port could run afoul of something accidentally and that would, quite literally, be the end of a major city. Accidents are one thing, and large numbers of people are trying to see to it that accidents do not happen, but we are not talking about accidents. Deliberate acts of sabotage or war against LNG tankers could be accomplished by an amateur in the art of sabotage. We rely for survival on goodwill.

The story of natural gas in this country parallels that of oil. It used to be considered a nuisance by-product of oil wells—you can still see photographs of oil wells aflame; they are burning off natural gas. It is now an important residential, commercial, and industrial fuel, as well as a favored fuel for electrical generation. It supplies one-third of the present U.S. energy demand. We have, in discovered reserves, anywhere between 2,000 trillion cubic feet and 375 trillion cubic feet of natural gas, about three-quarters of which comes from Texas and Louisiana. Production domestically has declined in the 1970s, and while we are the world's largest producer of natural gas, we are also the world's largest importer—most of this coming from Algeria.

Highly favored as a fuel, natural gas is non-polluting and it supplies U.S. industry with half its energy requirements. Most of it, both imported and domestic, flows through pipelines that originate in Louisiana and Texas. These, like oil pipelines, are largely underground and are therefore fairly impervious. In general, these underground systems are relatively invulnerable. As with oil fields, of which most natural

gas fields are a part, they are relatively dispersed and therefore relatively invulnerable.

Coal fields are widely dispersed as well. Coal is the most
plentiful fossil fuel in the United States, amounting to 88
percent of all the proven reserves of U.S. fuels. Thirty states
have mineable deposits, with Montana, Illinois, West Virginia, Pennsylvania, and Kentucky being the richest. The
West beckons greatly these days, having 70 percent of the
mineable coal in its bailiwick; most of this, however, is low
in energy value, high in sulphur content, and must be strip
mined. Nevertheless, environmental problems aside, the
western states are seeing the resurgence of boom towns—
boom regions—as this resource is extracted. On the other
hand, improvements in deep-mining techniques, such as the
development of the continuous mining machine, have
helped reduce the labor intensity of deep mining and so the
eastern coal mines are seeing a resurgence as well.

Even though coal is a plentiful—in fact, an underex-
ploited—resource the import litany holds true again. In 1978,
the United States imported three million tons, more than a
million tons more than we exported in the year before. Most
of these imports were from Australia and South Africa,
though some came from Canada, Poland, West Germany,
and the Netherlands. Coal travels a lot, most of it by rail and
lesser amounts in pipelines mixed with water or in trucks.
Except for the one-fourth of coal transportation that takes
place in pipelines, most of this transportation depends on
diesel fuel. Coal, as a fuel, is complex in its environmental
considerations and much affected by the needs here and ther
for different kinds of coal, but it seems relatively invulnerable as a system. Except, as the Congressional Research Service has pointed out, ". . . our dependence on overseas
sources for half of the oil we use may threaten our supply
lines for coal as well."

A closer-to-home limiting factor on the development of
our coal resources is water. The National Academy of

Figure 7. Coal Fields of the United States

Sciences, among others, has drawn attention to the fact that in the most coal-rich areas, the local supply of ground water or surface water is insufficient to meet the consumptive use of the coal producing process. And beyond the cooling and other processes of energy conversion, water is needed to rehabilitate strip mined areas. In short, the water-rich process of mining and converting coal to useful energy sources requires vast quantities of water, and most of our coal occurs in semi-arid regions of the West where the use of water is more controversial than any other subject. There is talk in the West of defying federal intervention (the sagebrush revolt), and there is talk of disenfranchising Indian tribes, who sit on vast energy and mineral deposits. None of this disputatious business can compare with the problems of the use and distribution of water. Nevertheless, coal appears to be the most useful, most plentiful, least vulnerable of conventional sources of energy available to us, and there are a number of unconventional uses being planned for it. These, as will be seen, have their own clouds hanging over them.

Metaphorically, one of the cloudiest sources of energy is

nuclear power. Currently, nuclear power contributes 10 percent of our electrical generation-needs and about 2 percent
of our total energy needs. On the other hand, it uses a vast
amount of our mental, emotional, and political energy. Some
states—notably Illinois, New York, Connecticut, Pennsylvania, South Carolina, Florida, and Alabama—rely on it for
substantial portions of their electricity. There are in all sev-
enty-one nuclear power plants in the country, with construction permits granted for ninety more. Another forty-five are
planned. It does not take much effort to transport nuclear
fuel (enriched uranium) to a nuclear plant, so most plants are
located as near to the market they serve as is possible, given
safety and environmental considerations—thus cutting down
on the entropic inefficiencies of transmitting electricity.
They are, therefore, highly centralized installations.

On the other hand, nuclear fuel processing plants require
a great deal of energy, metal, and water. In fact, they have
to be located alongside streams or other sources of water. So
do nuclear power plants, inasmuch as they call for large
quantities of cooling water. It is no accident that the Three
Mile Island plant is on an island.

There are a number of, by now, well-known problems
associated with nuclear power plants, the least publicly discussed of which is the decreasing availability of the needed
fuel: uranium. According to Warren Finch of the United
States Geological Survey, "The uranium found thus far was
easy. It was at or near the surface. The new ore for the future
will have to be found in deeper horizons and be of lower
grade." No major new uranium mining districts have been
found in the country in years.

Another series of problems afflicts nuclear power—its
high demand on energy, materials, and capital—and we shall
return to these later. For now, it is sufficient to point out that
we do import uranium and that these currently operating
nuclear power plants, like those of all other nations who have
them, are strategically important targets counting heavily in
the planning of warlords in the Pentagon as well as the

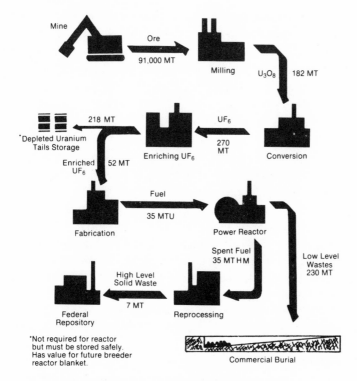

Average Annual Fuel Materials Requirements for a Typical
1000 Mwe Light Water Reactor

Figure 8. Diagram of Nuclear Fuel Cycle

Kremlin. Even the threat of an attack on nuclear plants, releasing, as it would, vast quantities of radionuclides into the environment, might be enough to render a small nation supine.

In *Destruction of Nuclear Energy Facilities in War,* Bennett Ramberg notes that nuclear facilities are also extremely vulnerable to sabotage. He argues, "Despite the fact that many installations incorporate security measures designed to impede or prevent access to facilities, these measures vary. At some facilities, security is limited to fences topped with barbed wire. . . . Although these measures provide protection against lightly armed, rather unsophisticated intruders, most studies suggest that facilities are vulnerable to destruction by anyone with technical competence, including adequately trained military units, assuming they penetrate plant external defenses." Ramberg's book highlights his proposal for "legal restraint"—a recognition, under international law, that nuclear facilities are exceptionally vulnerable and that nations must agree not to target them. This approach is not "the most reliable option to minimize the military threat," he says, but "it offers a relatively expeditious as well as inexpensive means to address the problem by establishing a standard of behavior where one does not now exist."

We add that as the centralization of all our energy systems increases—with no plans for back-up or redundancy—we essentially have no protection outside of such a call for restraint. Without taking a definite path toward more local self-reliance and decentralization, the essential vulnerability of our civilization increases daily.

RESOURCE WARS

Recently there was a wonderfully encouraging development in the American use of materials. It advances both the pursuit of sophisticated technological ways of living and the grass roots movements to save the endangered sperm whales.

Certain engines operating at high speeds require lubrication by a very particular kind of oil—that produced by sperm whales. This dependence on an endangered species was something of an embarassment to a nation seeking to take the lead in preserving endangered wildlife in general and, specifically, in the International Whaling Commission. Then a young New Zealand chemist working at the U.S. National Academy of Sciences began to beat the drum for a scruffy plant that grows in the American Southwest called jojoba. It produces an oil that only a sophisticated chemist can distinguish from sperm whale oil. Highspeed engines cannot make this distinction.

The jojoba plants, which grow erratically and widely in the southwestern deserts, can be cultivated. In addition, interest in jojoba and another plant named guayule, which produces a high quality natural rubber, is increasing in the southwestern United States and in Mexico as well. We may witness the development of an entirely new industry in the next few years that will provide a reliable source of income for dwellers in these arid regions—an industry that will eliminate at least one incentive for the destruction of whales and will minimize the waste of petroleum products as well. Unfortunately, developments such as these are isolated in the modern world.

A great number of modern technological systems—from aerospace and electrical equipment, to nuclear power and communications—operate at high speed and high temperatures, which requires special substances, mostly exotic metals. Most of these must be imported to the United States from countries that are often *more* unstable than the oil-producing nations. Nor is it only the United States that hangs precariously on the teat of these nations for substances known as strategic materials. Most European countries imported their supplies of titanium from Russia until, in 1980, the Russians decided to curtail exports of titanium to Europe. Was it because Europe was building too many military airplanes and using a key Russian strategic metal?

Table 3 shows the strategic-materials-reliance of the United States. The only major industrial material for which we do not import more than 50 percent of our consumption is iron ore. Most are in the 90 percent range—some at 100 percent. Exporting nations, along with a few others, might well become MPEC—indeed, it is surprising that they have not already done so. Furthermore, the Soviets have been establishing contacts and power bases in or around many of these regions, causing concern about the future availability of their minerals.

The United States has a stockpile of strategic minerals—

Table 3. U.S. Reliance on Strategic Materials

Material	Percentage of U.S. Consumption from Imported Materials
Titanium (rutile)	100[f]
Columbium	100[d]
Tin	100
Beryllium	100 (approx.)[b,c]
Germanium	100 (approx.)
Platinum	100 (approx.)
Manganese	98[c]
Tantalum	96[c,d]
Aluminum	93[a]
Chromium	90[a,c]
Cobalt	90[c]
Nickel	77[e]
Tungsten	59
Copper	n.a.
Molybdenum	n.a.

a: Reliance on politically unstable regions
b: The United States has large potential supply.
c: Reliance on politically unstable African region (e.g., South Africa)
d: Reliance on politically unstable Asian region (e.g., Thailand)
e: The United States has large potential resources, but domestic production has been limited due to technological and environmental problems.
f: Reserves have been identified in the United States, but none mined.

about $13 billion worth—but there are shortages and imbalances in several principal categories that would require an estimated $46 billion to bring the inventory to stated goals. The United States spends a tiny fraction of this to maintain stockpiles. Quotas have been set aside that mandate a monthly percentage of the materials to defense-rated orders; the remaining materials are available for market consumption—for example, the nuclear energy industry. A nuclear power plant uses such materials as aluminum, antimony, asbestos, beryllium, cadmium, chromium, cobalt, lead manganese, mercury, nickel, platinum, silver, tin, tungsten, and zinc. Many of these critical materials are used in reactor cores and, because of excessive radioactive contamination, cannot be recycled in the future—a contingency unknown to other energy technologies. In the event of an embargo, which a number of analysts see as a distinct possibility in the future, we might find ourselves bereft of key strategic materials because of our reliance on the nuclear fuel cycle.

In addition to energy wars, therefore, we also face resource wars. The National Strategic Information Center recently released a white paper urging that America adhere to "resource war" tactics, suggesting that "the U.S. should design new alliances and be prepared to intervene militarily, to be guaranteed access to Mideast oil and southern Africa minerals. . . ." One way to alleviate this particular vulnerability is to stimulate domestic production. Tax codes, antitrust legislation, and environmental regulations appear to make domestic production undesirable. It is, moreover, thought to be "cheaper" to get materials of this sort from nations with low labor costs. For example, chromite production in the United States ceased in 1961 because it became uneconomic. Another example is colombium production. The U.S. Bureau of Mines found little of this substance in the United States and estimated worldwide production could not exceed one million pounds per year. The price at the time was $1.60 a pound. When the government said it would pay a 100 percent bonus per pound, it took only three years to put fifteen million pounds in the stockpile. Policymakers are now learning to weigh national security against price.

On March 13, 1981, President Reagan announced the creation of a new fund to expand the nation's stockpiles; concurrently, one of Wall Street's largest firms, Bache Halsey Stuart Shields, Inc., announced that it would begin selling exotic metals to investors. Another firm, the James Sinclair Company, has placed chromium, cobalt, germanium, and manganese on its list of "primary investments," but this company also suggests purchases of antimony, indium, rhodium, and titanium.

The Reagan approach to the resource problem, as unveiled by Interior Secretary James Watt and Secretary of State Alexander Haig, is to open up former wilderness areas and substantial blocks of federal land to prospectors looking for new minerals. Another hot area for prospecting is the seabed; Reagan dismissed the entire U.S. delegation to the United Nations Law of the Sea conference due to his advisers' fears that a new conference treaty could jeopardize deep-sea mining by U.S. companies.

What has *not* been considered by President Reagan and his cabinet are some of the short-term, productive answers to the strategic materials dilemma. They include national policies to shift industries away from reliance on rare materials and toward substitute materials and processes. Encouraging lighter, more fuel-efficient vehicles, along with a greater emphasis on urban mass transit, would help shift needs away from materials-intensive industrial processes and away from energy and materials-intensive forms of transportation, such as jet travel. Encouraging new industrial processes, which would utilize less energy and materials, may be the quickest "fix" to the new fear of a resource war.

5. Centralized Energy Projects

CENTRALIZATION OF the U.S. energy system has occurred at an ever-increasing pace over the last century. Our current system, as we have seen, is energy- and materials-intensive, and the future appears cast as an extension of the past trend of centralized projects. Ranging from nuclear power plants to massive synfuels complexes, plans are underway by industry and government to accelerate the construction of ever more concentrated, central facilities.

THE ALCHEMY OF SYNFUELS

Acting quickly in the first weeks of Ronald Reagan's presidency, budget-cutters at the White House Office of Management and Budget (OMB), headed by David Stockman, tried to cut billions from one of the largest pork barrels in Washington's history. This was the Synthetic Fuels Corporation, a quasi-governmental, quasi-private organization designed to supply $88 billion in public subsidies to the U.S. synthetic fuels industry. Although President Reagan lopped off some of the public corporation's expenditures in the first round of his attempt to balance the federal budget, his administration weakened under intense pressure from the Congress and restored most, if not all, of the money designated for synfuels projects.

Synfuels, as they are called, are not synthetic fuels at all. They are naturally occurring materials like coal, which, with the application of additional energy, can be turned into a more convenient fuel, such as pipeline gas. On the other hand, the manufacture of synthetic rubber, which became the rage in World War II when supplies from the Asian rubber plantations were being cut off, is a truly synthetic affair. By using petroleum products, you can get a substitute for rubber and no longer have to rely on the vagaries of the nations that supply natural rubber. It is reassuring. Synthetic fuels, as a rubric—and one that has been used since the early part of this century—is a reassuring phrase. With synthetic fuels, the theory goes, we have the opportunity to solve the energy crisis and the increasingly well-perceived national vulnerability arising from dependence on other nations for our fuels. Maybe, maybe not. Without dwelling on semantics too much, it is useful to keep in mind that there is a lot of jargon in the energy business that can, in subtle ways, mislead the anxious and unwary. We will go along with the misnomer because there are, in fact, far greater problems with synfuels than their name.

After a large number of fits and starts over the past eighty years, United States energy industries and the U.S. Government have made a major commitment to synfuels as an energy source. The Energy Security Act of 1978 called for the synthetic fuel equivalent of 500,000 barrels of oil a day by 1987 and two million barrels a day by 1992. Upon signing the bill, President Carter said with pride that this action would "dwarf" the combined Marshall Plan, moon program, and interstate highway system as public works projects. Two million barrels of oil per day amounts to less than one-third of what we import *today.*

Nonetheless, even given President Reagan's budget axe, the U.S. synfuels effort will continue to gain steam during the 1980s, and a whole new range of favorable government "hidden" subsidies—such as the relaxing of controls on Western public lands and development payments through the federal Interior and Energy Departments—will reward this industry. In the meantime, taxpayers are allowed only a measly 15 percent credit on a few items of conservation technology—such as weatherstripping for doors and windows and insulation for the roof. A real program of national conservation, as we shall see later in the book, could result in saving our nation from the clutches of OPEC during the 1980s, rather than waiting decades for over-priced fuels from the new synfuels industry.

The two million barrels per day of the equivalent of oil that the synfuels investment will provide is slated to include only 500,000 barrels per day of actual oil substitute. To understand why this is so requires an explanation of how synfuels are made and where they come from.

At one point in their journeys through western America, Lewis and Clark paused for the night and lit a campfire against a cliff. Before long, the story goes, the cliff began to smoulder. It was an oil-shale cliff. This rock, which is technically not shale, occurs in various places around the country, but nowhere in such bounteousness as in the region where Colorado, Utah, and Wyoming meet and where the promising geology is called the Green River Formation. The rock contains a substance called kerogen. When the rock is melted, kerogen runs out in a liquid form not unlike crude oil. Thenceforward, in fact, the matter of turning kerogen into useful petroleum products is not much more complex than refining regular oil. And in the Green River Formation alone, it is estimated, there is a supply of kerogen awaiting extraction that would last a hundred years.

There are a few problems with oil shale, however, one of which is the popcorn effect. If you heat up oil shale, it gains in volume. Therefore, once you have dug up a seam of the stuff, you cannot simply put it back. Some 25 percent of the tailings has to go somewhere else. Another problem is that it takes a great deal of energy to mine and process oil shale, not to mention the energy it takes to stow the detritus and ship the product to its destination. One estimate is that for every barrel of oil-shale oil we produce, we will have to expend the equivalent of one and a quarter barrels of oil. In most forms of accounting, this is known as a net loss.

It may be that those 500,000 barrels a day by 1990 may be well worth the net energy loss to produce them. Strategic considerations may, as for the Germans in World War II, make it a useful expenditure. However, it is evidently not one that any nation can make indefinitely. The *dollar* price of this oil will also surely be high—indeed, the Energy Security Act's provisions make certain that, were oil-shale synoil to become relatively inexpensive to make as a result of some technological breakthrough, there will be price supports—keyed to the going rate of imported and domestic oil—to make sure that synoil prices remain equally high. But ten years seems like a perilously long time to create, at vast expenditure of capital, manpower, and resources, not to mention environmental complications, what amounts to one-sixteenth of our *present* imports of oil. This, on its face, would seem to be a relatively inefficient way to reduce the national vulnerability we suffer at the hands of energy-exporting nations.

The other three-sixteenths of our present import requirements that this history-dwarfing public-works effort will save us by 1992—and this is assuming that there are no delays, no slip-ups, no erroneous estimates based on an untried technology—will come largely from the processing of coal into other forms of fuel. Government and industry are also researching ways of using biomass to make fuel (that is, wood chips, agricultural waste, and the like) and are looking at solar technologies, but the amounts of dollars thus allocated are a vanishingly small fraction of the big nut.

Coal is truly the stuff of the new American dream. In chapter 4 we saw how much of it there is—a supply that could last us for generations. We could become OPEC all by ourselves (and don't think that hasn't been gossamered up in a reassuring way by some energy pundits). Coal is not only plentiful in America, it is also marvelously malleable. A product of the dying of plants long ago in geological time, there is no end of things you can make out of coal. For example, given available technology you can take coal out of the ground, put it through a process called gasification, and come out at the other end with a substance called medium BTU gas. If you take medium BTU gas and run it through a process that cleans impurities out of it and otherwise refines it, you have something that resembles natural gas. It is called synthetic natural gas, or SNG, and it can run your stove or your local electricity generating plant. Or it can be converted into methanol, which can run a car. Synthetic natural gas is a methane-rich gas from which many impurities have been removed. Its heating value is about 1000 BTUs per standard cubic foot, and it can be produced from coal, agricultural and forestry wastes, municipal solid wastes, and a range of other organic materials. It will, as planned, be produced principally by gasifying coal, which will normally be done at the mouth of the mine.

Medium BTU gas, is produced from coal that is crushed and put into large pressurized reactors, where it is heated up above 1500 degrees Fahrenheit and mixed with steam and oxygen. This produces a mixture of hydrogen and carbon monoxide—a gas that has about one-third the heating value of natural gas. Some methane is also produced. Medium BTU gas is what used to be called town gas and it lit U.S. city streets and cooked American meals until as recently as 1947. It still can serve as an industrial fuel since it is easier to use than coal and is the only other form of energy available to industries which cannot burn oil or natural gas in new boilers. There may be other uses for medium BTU gas around the corner. It could be further processed into methanol, which has half the energy content of gasoline; the Mobil Oil Company, with federal support, is exploring ways to transform it directly into gasoline. All of these products require a considerable amount of energy and money to be produced from the stage of medium BTU gas—as does the manufacture of synthetic natural gas. To accomplish this, the medium BTU gas must have its hydrogen content nearly doubled and be reacted catalytically to produce methane, the chief ingredient of natural gas, which must then be dried.

Recent estimates project the cost of such SNG between $5 and $8 per million BTUs (in 1979 dollars) as opposed to $4.65 for Canadian natural gas and less for domestic. Natural gas prices will certainly rise, though to what extent no one is sure. It is hoped, however, that synthetic natural gas will be "marginally" competitive with the natural product, but this hope is based on a lot of assumptions that cannot be proven without commercialization. If some current efforts in the synfuels area are indicative, then we can expect to find that the assumptions will be proven wrong. For example, in 1978, the Department of Energy sponsored the building of a plant in West Virginia to make liquid fuels from coal, selecting the Pittsburgh and Midway Coal Mining Company to design the plant. By December, 1980, the project had fallen fifteen months behind schedule and, though it was still in the design stage, the projected cost had risen from an original $700 million to about $2 billion. Indeed, Thomas N. Bethell, an editor for the *Washington Monthly* has compared synfuels to the Bermuda Triangle: "everybody wants to believe it doesn't exist, but projects keep disappearing into it just the same."

There are, of course, other problems to be faced concerning the question of synfuel production besides its engineering and its economic feasibility. Environmental problems abound. It takes about four gallons of water to produce one gallon of fuel by these processes, and most of the synfuel plants will be located where water is scarce. Furthermore, the water so used is not recyclable. Most of these areas are also now used largely for agriculture. The full development of coal resources in twenty-four states, it has recently been estimated, would take nearly 50 million acres of prime farmland out of circulation; it is yet to be proven, moreover, that they could be restored to anything much more productive than scrub pasture land. But one thing seems certain: economies of scale will factor in such operations. You can make low BTU gas in relatively small plants, but for medium BTU gas and for synthetic natural gas, production installations must be large in order to achieve even marginal economic effectiveness—on the order of a 500-megawatt power plant. Thus, this particular energy panacea, while being commercially "iffy" also contributes to the strategic problems of a centralized national energy system, chiefly making the system yet more vulnerable.

There is nothing wrong, *per se,* with making useful fuel from a variety products—that is, synthetic fuels. The problem we have with synfuels as currently conceived is that the most inefficient and expensive ways to produce them have been selected and subsidized. We have to wonder why, if the manufacture of synthetic oil from oil shale or synthetic natural gas from coal is so attractive a process, the energy companies did not see fit to go into the business until the federal government waived the subsidy banner. It seems a fairly harsh dialectic whereby the government extracts money from the middle class and passes it along to what might be called the tax-free superclass. They in turn, while operating at no risk and posing as capitalists, complain of government interference and eventually raise the price of the product they sell to the middle class, who paid the bill in the first place. Resentment, in such a situation, is a reasonable civic emotion. There are a number of rather unhappy examples of how this new American dialectic has worked in the past, one of which follows.

"WHOOPS"

Citizens in the State of Washington refer to the Washington Public Power Supply System with their own acronym: Whoops. The WPPSS undertook in the early 1970s to construct the nation's most ambitious nuclear generating facility

—five 1,000-megawatt generators near the city of Richland, not far from the federal government's Hanford nuclear reservation. Richland would become, vernacularly, Atomsville, USA. The project, it was estimated, would cost $4 billion, financed by issuing tax-exempt bonds. The array of generators would be able to supply enough electricity to satisfy the needs of five cities the size of Seattle, or, 2.5 million people. It would be on these plants that the future economic growth and well-being of the state would thrive. The tax-exempt bonds sold like hotcakes, particularly because three of the five plants were backed by a remarkable arrangement wherein the federal government's Bonneville Power Administration, which distributes electricity to 6.5 million people in the northwest, would raise its rates as high as needed to cover the costs.

To achieve economies of time, the construction proceeded with the design. In other words, each stage of the five plants, each a bit different from the others, were to be completed as or before the next steps were being designed. The result, as one might expect, is that the estimated cost is now $17 billion and still growing. The Nuclear Regulatory Commission has regularly found defects—in one case, the steel and concrete shield around the reactor was judged to be faulty—and the first plant is six years behind schedule. Not surprisingly, WPPSS is finding it difficult to market new tax-exempt bonds to finance the spiraling costs of this Brobdingnagian project, which one observer compared to the idea of building all of the pyramids at the same time. Not only the local residents refer to the WPPSS as Whoops, so do, according to the *Wall Street Journal,* the bond traders of New York.

Recently, a former federal energy official, Robert L. Ferguson, was brought in to save the project. His comment was that, "If we cannot do it here, nuclear may very well not make it in this country." He pins his hopes on the fact that even with all the cost overruns, the price of electricity in Washington will still be less than the national average, because it will be melded into a system that uses a great deal of very inexpensive hydroelectric power. The citizens of what would have been Atomsville and 6.5 million other rate-paying hostages are not pleased—Nor are the bond traders, and neither is the Government Accounting Office, according to which the $4 billion projected as necessary for completing the last two plants would be better spent on energy conservation and renewable energy sources.

Meanwhile, Whoops, like its brethren in the nuclear industry, stands out as a prime example of the end result of a $30 billion subsidy to the nuclear industry since World War II. As the utility industry (both public and private) decries environmentalists and anti-growthers, the industry fails to make commitments for far more efficient power plant investments, even including smaller, less costly nuclear facilities. Instead, the industry has opted for the 1000-Megawatt-class nuclear pyramids, designed by engineering consulting firms that operate from manuals approved by the federal government. Taking subsidies for this process in stride, the industry is asking for still more federal handouts, while the Wall Street bankers hide. And meanwhile, an anxious citizenry is told that accidents such as the one at Three Mile Island, for which they are asked to fork over the repair costs, prove the safety of the system—and they wait for the other shoe to drop.

THE STRATEGIC PETROLEUM TARGET

Ben Franklin would have approved. In fact, who can gainsay the notion of setting aside a little reserve for a rainy day? In 1976, Congress, recognizing that a sudden cut off of oil supplies from abroad for whatever reason would be disastrous, passed legislation that called for the construction of a Strategic Petroleum Reserve. It was, by 1980, to contain 500 million barrels of oil—a supply that would nourish the nation for several months in the event of any untoward occurence—and ultimately a billion barrels.

The project has turned out to be a considerable embarrassment. By the end of 1980, the SPR was only 20 percent full. The decision had been made for reasons of price not to store this oil in aboveground containers, but instead to put it into underground "storage tanks" in naturally formed salt domes in Texas and Louisiana. The government proceeded to buy up old salt mines in those states and to make them a suitable repository for oil, as well as to create new cavities. The process is as follows: fresh water or seawater is pumped into the salt dome to dissolve the salt. The brine is piped several miles out into the Gulf of Mexico, and the crude oil is pumped into the cavern. Then, in the event of an emergency, this plan suggests, you simply pump oil out of the salt cavity by pumping in water, cleanse it of salt, and ship it off to where it is needed. The process seems straightforward enough. There are, however, a few problems that have developed, along with some that apparently no one has noticed yet.

It takes seven barrels of water to dissolve the salt for each barrel of oil to be stored. Each cavity would be big enough to hold ten million barrels of oil. No such effort had ever been undertaken before. It ran into immediate engineering snags, as well as arguments from shrimp fishermen in the Gulf of Mexico who did not want a lot of unnatural water poured into their shrimp beds. The result was longer pipelines into the Gulf—and numerous delays. When, in the late seventies, the oil began to flow toward the SPR, the SPR was not yet ready. The Office of Management and Budget vetoed a recommendation that the oil be stowed temporarily in tankers and other aboveground storage facilities. Just about that time, the Iranian revolution took place and the administration decided not to make any more purchases of oil for the SPR. Then, as the oil market eased in the spring of 1980, Saudi Arabia voiced its objections to the concept of the SPR. The administration resumed filling the salt cavities, now ready to receive 250 million barrels. So now, after six years, the SPR contains

20 percent of the origin 500 million barrels called for—a kind of Whoops II.

In a recent book *Energy and Security* David A. Deese and Joseph S. Nye have made a strong argument for the acceleration of the fill rate of the SPR. And it is hard to dispute the idea that we should act promptly to produce an emergency reserve. They say that such a reserve would "preserve a healthy economy during interruptions, nurture cohesion in the alliance, and gain leverage over producers. . . . It is vitally important to get the fill rate up to a reasonable plateau quickly." At the present rate, they point out, it would take until 1991 to reach the sought after goal of one billion barrels. Meanwhile, the Department of Energy has estimated that we really need something like three or four billion gallons.

Deese and Nye point out that private reserves "ironically" exceed the amount currently in the SPR; they total approximately 1.36 billion barrels, or, an equivalent of 194 days of imports. Several European nations rely upon the oil industry itself to maintain stockpiles against an emergency, with certain minimums mandated by law. The U.S. oil industry disapproves of such ideas and does not want to rely on government judgment about the necessary amounts to be stored. Given the government's own record in this matter, one can hardly blame the oil companies. But one can also wonder (along with *Fortune* magazine), "[if] the oil industry can't be jawboned into hanging on to all those barrels, perhaps Washington should offer to defray some of the carrying costs." Why does there need to be any subsidy at all to encourage the oil industry to act like Aesop's ants—in their own interest? Those 1.36 billion barrels will be saleable at whatever the going price might be at the time of their release to consumers.

Quite aside from the issue of who should be responsible for stowing oil, there is a geographical—and ultimately strategic—consideration: why stow it in Texas and Louisiana? What this means, broadly stated, is that we pump oil in

Alaska, say, and ship it the length of the continent and pump it into a briny hole. Then, if an emergency occurs, we pump it out of the hole (at a rate of fifteen million gallons a day, it is claimed), purify it, and ship it halfway back the length of the continent to use in Illinois, or 1500 miles to California—where much of it would have originated in the first place. Somehow, in all of this, inefficiency has been elevated to the level of inexplicable incompetence.

Furthermore, to locate the nation's oil insurance policy in one place, even if it is down a few thousand feet underground, is to stick one's jaw out in the manner of Mohammed Ali (who lost) and beg for an act of sabotage. It is to paint a bull's-eye around your future for the benefit of your adversaries.

THE PROLIFERATION OF HIGH-GRADE DANGER

During the early post–World War II years, many citizens feared that producing electricity from nuclear energy meant there would be an atomic bomb in the plant that could go off at any minute. Part of the U.S. government's multibillion-dollar job in launching the nuclear power industry in this country was to overcome such naïve concerns about technology. Most people now know perfectly well that there is not a bomb in their neighborhood nuclear power plant down by the river. They are relieved to know that the engineering principles for a bomb and for a power plant simply are not the same. And they also happen to be—ultimately—wrong. Every week your local neighborhood nuclear power plant produces just about enough plutonium to make a bomb. So far, worldwide civilian nuclear power plants have produced 100,000 kilograms of plutonium. And a team of high school students good enough to win the annual science fair awards from the American Association for the Advancement of Science would be able to figure out how to build a nuclear device ("device" is governmentese for "bomb") if only they could get a little capital and some of that plutonium.

The difficulty in getting that plutonium is that it must be separated out from the other nuclear wastes generated by the reactor and then purified. This is called reprocessing. Reprocessing of nuclear fuels is not being done in this country at the present time, given the possibility of sabotage and the diversion of nuclear materials. Nuclear wastes, however, are being processed elsewhere. Until recently, U.S. nuclear wastes were being processed in several European nations, after being shipped there at a profit to the U.S. nuclear industry. And, over the years, the U.S. government has provided other countries with the technical know-how to produce nuclear-generated electricity and to process it into plutonium. Among the nations that are now able to do this is our reliable friend, India. How did all this come about?

Early in the development of the nuclear industry in this country, it was recognized in a few government circles that there were some dangerous aspects of the process, chiefly the production of weapons-grade plutonium from nuclear wastes. David Lilienthal, who headed the Tennessee Valley Authority and was later first chairman of the AEC (and who, before he died in 1981, was a proponent of both small hydroelectic plants and revised safety standards for the nuclear industry) had co-authored a report pointing to these dangers. He urged that the processing of nuclear wastes be internationalized—that is, put under no individual country's control. The Eisenhower administration began the Atoms for Peace Program and made attempts to get international treaties for the control of these processes. Somehow in the procedures that followed, the word "dangerous" was replaced by the word "sensitive"—and somehow it all got out of control.

As Victor Gilinsky, a member of the Nuclear Regulatory Commission, said in a speech to the League of Women Voters in late 1980 in Maryland, "The idea of international safeguards as a means to prevent ready access to dangerous materials thus deteriorated into a system of inspection and measurement with no clear rationale, except perhaps to drop the cloak of legitimacy over nuclear commerce."

In the late 1950s, the United States made plutonium-separating technology widely available. And around that time we began to dream of the nuclear breeder reactor, a device that would extract plutonium by reprocessing spent nuclear fuel and, if uranium was expensive and reprocessing cheap, become something of a nuclear perpetual machine. We were headed for a plutonium economy. However, in the 1960s it became clear that sending little packets of plutonium here and there by train, plane, and ship was dangerous. It also became clear that uranium was relatively cheap and reprocessing was rather expensive. A Nonproliferation Treaty was proposed and ultimately signed—over the howls of despair on the part of the nuclear energy industry here and abroad. Germany squabbled for a year to protect its commercial advantages in manufacturing plutonium. India finally refused to sign—quite properly, since India was at the very moment working on what it liked to call a "peaceful" nuclear device, which it subsequently exploded to the shock of the world. Soon France announced plans to sell a reprocessing plant to Pakistan, and the Germans, it was learned, were selling reprocessing and enrichment technology to Brazil.

Subsequently, President Ford called on the U.S. industry to cease all commercial reprocessing and called on the nations of the world to do the same until adequate safeguards could be imposed. His successor, President Carter, endorsed this policy; the British and French, however, thinking that we were not really serious, went on with their own billion-dollar plans. The Germans and the Swiss believed that U.S. efforts for nonproliferation were merely an attempt to gain an advantage for the U.S. industry. In any case, the U.S. Congress passed a more stringent Nuclear Nonproliferation Act in 1978—much opposed by that child of government, the nuclear industry—that tightened up even further on nuclear exports. No nation, it said, that refuses to accept international inspection can receive nuclear materials from U.S. exporters. Just how tight this legislation was became clear two years later when the Carter administration, with the opposition of the House of Representatives but with the consent of the Senate, reversed itself and accepted an application from India for more nuclear fuel.

The nuclear industry has plenty of reason to worry about any such controls on its activities and will certainly soon launch a new attack on legislative or regulatory controls on its exports of fuel and other technological activities. (And that is largely because the growth in demand for electricity in this country has dropped.) Also, as Gilinsky said, the industry "is stuck with a terrific overcapacity and is desperate for orders." There have been few, if any, orders in the past few years for the building of nuclear power plants. So one of the only ways the industry can make any money is from exports.

As Commissioner Gilinsky pointed out, "Nuclear controls have never had any constituency but the old-fashioned, common-sense fear of blowing up the the world."

6. Henry Adams's Physicist

—Henry Adams, 1910
"A Letter to American Teachers of History"

"From the physicist's point of view, Man, as a conscious and constant, single, natural force, seems to have no function except that of dissipating or degrading energy. Indeed, the evolutionist himself has complained, and is still complaining in accents which grow shriller every day, that man does more to dissipate and waste nature's economies than all the rest of animal or vegetable life has ever done to save them. 'Already,'—one may hear the physicists aver—'man dissipates every year all the heat stored in a thousand million tons of coal which nature herself cannot now replace, and he does this only in order to convert some 10 or 15 percent of it into mechanical energy immediately wasted on his transient and commonly purposeless objects. He draws great reservoirs of coal-oil and gas out of the earth, which he consumes like the coal. He is digging out even the peat-bogs in order to consume them as heat. He has largely deforested the planet, and hastened its desiccation. He seizes all the zinc and whatever other minerals he can burn, or which he can convert into other forms of energy, and dissipate into space. His consumption of oxygen would be proportionate to his waste of heat, and according to Kelvin, "If we burn up our fuel supplies so fast, the oxygen of the air may become exhausted, and that exhaustion might come about in four or five centuries." He startles and shocks even himself, in his rational moments, by his extravagance, as in his armies and armaments which are made avowedly for no other purpose than to dissipate or degrade energy, or annihilate it as in the destruction of life, on a scale that rivals operations of nature. What is still more curious, his chief pleasures, so far as they are his own invention, consist in gratifying the same unintelligent passion for dissipating or degrading energy, as in drinking alcohol, or burning fireworks, or firing cannon, or illuminating cities, or deafening them by senseless noises. Worse than all, such is his instinct of destruction that he systematically exterminates or degrades all the larger forms of animal life in which nature stored her last creative efforts, while he breeds artificially, at great expense of his own energies, and at cost of the phosphorus and lime accumulated by nature's mostly extinct organisms, the feebler forms of animal and vegetable energies needed to make good the prodigious waste of his own. Physicists and physiologists equally complain of these tendencies in man, and a large part of their effort is now devoted to correcting them; but the physicist adds that, compared with this enormous mass of nature's economies which man dissipates every year in rapid progression, the little he captures from the sun, directly or indirectly, as heat-rays, or water-power, or wind-power, is trifling, and the portion that he restores to higher intensities would be insignificant in any case, even if he did not instantly degrade and dissipate it again for some momentary use."

Part Two

AVAILABLE TOOLS

The care of human life and happiness, and not
their destruction, is the first and only legiti-
mate object of good government.

—Thomas Jefferson
1809

7. Efficiency and Energy

I<small>N THE</small> preceding chapters of this book, we have looked at some of the weaknesses in the American energy system—including the centralization of the system itself and the problems created by the spiraling demands on scarce fuels and minerals. The conventional political wisdom in American society seems to dictate that the only salvation from our vulnerability dilemma can come from building more power plants, refineries, and other energy facilities. We believe that a new technological revolution is dawning, one that will focus on creating entirely new industries in America that will use less energy and resources, employ more people than the capital-intensive industries of today, and create new wealth.

We have only to look at the intensive changes in technology brought about by developments in microelectronics to get a glimpse of this future. When the Electrical Numerical Integrator and Computer (ENIAC) was built at the University of Pennsylvania in 1946, ushering in the computer age, it used enough energy to operate a locomotive and could do about five thousand calculations per second. Today, fast computers can perform over eighty million calculations per second; even simple hand calculators—that outperform ENIAC—use an insignificant quantity of energy from tiny batteries. A great opportunity for our industrial civilization is to find like ways to replace the brute use of energy with technologies that conserve resources and increase economic output. We believe not only that this goal is achievable, but also that the technologies that can make it happen are in our midst today. In the following chapters, we will look at some of them.

In the years since the Arab oil embargo shocked the West, a veritable bombardment of information about the "energy crisis" and ways to solve it has filled the airwaves, magazines, and newspapers. One of the most popular approaches has been the idea of energy conservation, of finding ways to shift our uses of energy away from wasteful patterns. Yet, much of this message has been misinterpreted by political and economic leaders as a call for ending the "good life." When President Reagan took office, one of his first official actions was to end the federal government's mandatory building energy standards, which had set temperature levels in public buildings for seasonal energy conservation. The president quipped that energy conservation was essentially "staying hotter in summer and colder in winter."

We take issue with this notion. Energy conservation does not mean deprivation. It means economic readjustment to meet the resource realities of this last quarter of the twentieth century. During the past few decades the United States has vastly increased its energy consumption, while many European nations and Japan have consistently pursued economic and industrial policies toward more efficient housing stock, transportation systems, and industrial approaches—all in order to make better use of resources and energy. As a result, the economic "threats" from Japan and Germany are seen as a dire challenge to the future of key U.S. industries, such as the automobile industry. Scores of government officials and corporate executives are calling for protectionist legislation to cut down imports of foreign cars, which are twice as energy efficient as the guzzlers emitted from the decaying assembly lines of Detroit. Such policies do not recognize the economic trap implicit in such a shortsighted plan. Were the United States to engage in energy protectionism, we might soon find that the retaliatory measures by our trading partners would hurt us more economically than limiting car imports to protect Detroit would help us. German political leaders are already threatening the spectre of an international trade war, if the United States moves in this direction.

In the meantime, the prospects for economic revitalization through the use of new, highly efficient technologies looms bright as an economic growth area for the U.S. economy—in fact, it is the only economic growth sector that would offer the added benefit of decreased resource use. Providing new markets for energy-efficient technologies and renewable community-based technologies is an exciting challenge for the nation. A recent conference of industry and government leaders, held by the Aspen Institute, noted the prospects for such a decentralized beginning. The conference proceedings concluded:

> The potential for decentralized technologies as fuel savers or displacers in the electrical sector in the next twenty to thirty years is high—up to 20–25 percent of future generating capacity. These technologies include principally the solar ones (thermal, photovoltaic, wind machines, hydro); conservation technologies such as heat pumps, new appliances, and insulation; and cogeneration and fuel cells using fossil fuels. These technologies, especially the solar ones, are highly capital, materials, and energy intensive during the build-up time of their deployment, and so their benefits need to be discounted at least over 20- to 30-year time periods. Also, a production base for decentralized technologies needs to be established, and their equitable treatment in the rate structure needs to be formulated. Current research and development activities funded by the government, not-for-profits, and industry, provide a spectrum of innovative opportunities. The problem is to demonstrate that these technologies can provide economic and reliable service on the scale needed by users. How to finance these opera-

tional demonstrations is an open question: a proper balance of government, private, and ratepayer investments needs to be formulated. The goal should be to provide users with a wide range of true economic alternatives from which they can select the technologies of greatest utility to them, subject to governmental policies and regulations on rates, the environment, fuel use, and the health and safety of the public. The process of choice among these technologies, and of their demonstration, is the determinate question, rather than the establishment of specific end results on an *a priori* basis.

The first priority in establishing a more efficient economy is to encourage thousands of energy-conserving measures affecting individuals, businesses, and communities. Yet selling the idea of conservation seems to run counter to the prevailing wisdom and to the grain of America's economic history. Washington energy lawyer and former Joint Economic Committee energy specialist Jerry Brady says that conservation is misunderstood. "It seems downright un-American," he says. "Beside the Paul Bunyan thrill of picking up the surface of Wyoming, shoveling out the coal and putting the top back on, conservation is Charlie Brown doing his homework. Worse, to most people, conservation means hard times and high prices. It is true that high prices partly explain Europe's commanding efficiency lead, but we have missed the central point. Conservation flourishes in an open market, where it competes for capital on equal terms with production."

Today, the business of energy conservation is rapidly becoming a major industry. Roger Sant, formerly national director of energy conservation under President Ford and currently director of the Mellon Institute's Energy Productivity Center, thinks that today's marketers of electricity, oil, gas, and coal may well become tomorrow's "energy wholesalers." "Their retail role," he says, "will be taken over by new marketers of the end uses or services of energy—light, heat and motion instead of kilowatt-hours, therms and gallons.

And that change will bring some $400 billion in new markets for products and services not now considered part of the energy sector." Sant adds that "While there is some question about the future requirements for energy, there is little question about the requirements for these energy services. They will grow roughly in proportion to economic activity."

The Energy Productivity Center has pioneered the development of the "least-cost-strategy"—an approach based on the idea of supplying energy services and efficient ways of using energy in homes, industry, and transportation. In *The Least-Cost-Energy Strategy* Sant argues that "The nation has been on an unproductive course; it has looked at energy as a diminishing domestic commodity instead of a service which could unleash new competitive forces, multiply consumer choices, force down the real cost of energy services over the long run, cut the demand for energy supplies, and make a substantial contribution to the nation's economic health." The Center tested a hypothetical model of the U.S. energy economy, in which the least-cost approach was used —purchases of more efficient buildings, vehicles, industrial equipment, and so forth. The results, according to the Center, were "more than encouraging," and the results showed that a 10–15 year pattern of more energy-efficient purchases in the overall economy would have reduced consumer costs by 17 percent, while positively affecting the nation's economy and security.

BEGINNING AT HOME

Approaches to make the nation more energy-efficient must inevitably begin at home. And the homes of America indeed consume a great deal of energy, not including the use of the ubiquitous automobile and its very special fuel needs. About one-fifth of all the energy used in this country is spent in private residences. About half of that is spent for space heating, and the rest is divided up among a number of appliances, ranging from air-conditioners to refrigerators to water heaters.

In recent years, several factors have precluded consumers from doing much about the growing use of energy in the home. One of the major factors has been the artifically subsidized cost of energy by the government. In almost all areas, from natural-gas pricing policies to the rates set for electric utilities, energy markets have been controlled markets. Now that most of our energy sources have been "decontrolled" (with the glaring exception of politically favored sources such as nuclear plants), higher prices should serve as a stimulus for household conservation. Other factors, though, may work against this; not the least problem is the overall health of the economy itself. After all, people have to be able to afford conservation.

During the writing of this book, we tested a new, innovative device for household energy conservation and derived a few observations on the relationships between economics, the government, and conservation. The device is a small heat pump that operates in the basement of our home in northern Virginia. The heat pump is used, not for heating household space—the best known heat pump technology—but for heating water. It works in the same way that a refrigerator does, except that in a refrigerator, heat is "pumped" from the interior of the box into the room. In the water-heating heat pump, ambient air is taken from the basement; the heat in the air is extracted by a compressor in the machine's refrigeration circuit and then condensed to release heat to the conventional hot water tank. This little suitcased-sized machine can be purchased for less than eight hundred dollars, and it will cut the average family's electric hot water bill in half. This means that, in most areas of the country the heat pump will pay for itself in three or four years in energy savings. In savings to the country, the reduction in energy use could be quite dramatic—given the fact that the potential market consists of about 30 million domestic electric water heaters.

The water-heating heat pump we used was built by E-

Tech, which is a small Atlanta manufacturer that was started by an enterprising engineer and businessman named Glen Robinson. Robinson's interests in heat pumps was triggered in the early 1970s, when his former company, Scientific-Atlanta, engaged in a series of experimental tests on houses using solar collectors with heat pumps. In the tests, it was discovered that the heat pumps—which typically save 50 percent of the energy wasted in electric resistance heating systems—were so efficient that there was no compelling reason (at least, in moderate climates) to add the expensive solar collectors. Intrigued by this finding, Robinson discovered that the specialized water-heating heat pumps had been developed in the early 1950s, but no appreciable market had developed, because the cost of electricity had been cheap at the time.

E-Tech was founded to build and market the water-heating heat pumps, but the company soon found that breaking into the national market was not easy. One of the reasons for this was the fact that the federal government, and many states, were lavish in rewarding tax credits and subsidies for such things as solar water heaters, which cost several thousand dollars per house, but they provided no similar aid that would stimulate consumer interest in the new heat pump. In fact, the government's conservation tax credit form, which gives consumers a 15 percent tax credit for several conservation items, specifically *excludes* heat pumps of all types.

We thought it was strange that the government would choose such a course, since one of the major government-funded household energy conservation studies published for the Department of Energy specifically noted that water-heating heat pumps were a far better consumer alternative than solar collectors—when each system was compared against inefficient electrical resistance water heaters. The study was published in 1978 by Dennis O'Neal, Janet Carney, and Eric Hirst, of the Oak Ridge National Laboratory. It projected the energy and economic savings of using these devices to the year 2000, computing how fast they might penetrate the market and throwing in such considerations as increases in fuel prices. The use of these little heat pumps, they found, would produce cumulative national energy savings of 1.5 quadrillion BTUs (the equivalent of 260 million barrels of oil) by the year 2000 and would save $640 million for the households involved. On the other hand, the commercialization of solar hot water heaters would result in national energy savings of .7 quadrillion BTUs (less than half) in the same period. The economic savings to solar households would be only $383 million. But the government, by giving tax credits to people who install solar hot water heaters, would lose $345 million in revenues so the actual net economic savings to the nation would be an insignificant $38 million. It was a small point, perhaps, and hardly what you would expect to be offered up as a symbol for the solution of so ominous an array of problems as an energy crisis and an overwhelming national vulnerability problem. What kind of solution is a little heat pump?

No one really likes to think small, or to imagine using less rather than more. We are not a nation of Thoreaus, the woodstove boom to the contrary notwithstanding. But if more American citizens were to do some of the basic things to improve household energy efficiency—like using storm windows, insulation, and more effective heat pumps—less energy would be needed in the central power grids, and a decrease in imports would vastly help our nation's security. In fact, on a local level, changes in energy use are already increasing, and the power utilities are sensing the need for change as well.

THE TREND TO SMALLER POWER PLANTS

Certain thinkers, among them some energy economists, are beginning to perceive not just the disadvantages of the larger power plants but the advantages of smaller-scaled ones. More than a decade ago, a 5000 megawatt, coal-fired generating plant was proposed for the Kaiparowitz Plateau, a region in southern Utah where not many people lived, where there was plenty of coal around, and where—it was estimated—a lot of power could easily be shipped to the burgeoning population of the American Southwest. A problem occurred. Surrounding the Kaiparowitz Plateau were a number of national parks; people in the area had noted that the clear skies of the Southwest were already nearly opaque due to emissions from a couple of other large power plants in the Four Corners region and they defended the parks fiercely. Even Robert Redford joined the cause, and, while history does not indicate how instrumental this powerful luminary was, the Kaiparowitz project was dropped.

History does, however, show that a study was made in connection with this project that suggested that big coal-fired power plants are not very efficient—either as energy-producers or dollar-users. Andrew Ford and Irving Yabroff (respectively associated with Los Alamos Scientific Laboratory and SRI International think tank) looked into the differences involved in capital and operating costs of nine 250-megawatt units and four 750-megawatt units. The latter would produce 3000 megawatts; the former, only 2250 megawatts. Yet the two researchers found that the nine smaller plants produced the "same effective addition in system capacity as the four 750 megawatt units." Why? It has to do with reserve margins. Every power plant, just like every automobile or dishwasher, has to be shut down for a certain period of time for repairs. To account for accidents (in addition to anticipated repairs) all generating systems have a reserve margin—usually 15 to 25 percent above that of the peak load expected from the system. Larger plants, history shows, have a higher "forced outage rate." If one large plant goes out it means you have got a lot of electricity to replace.

Ford and Yabroff concluded that smaller plants have a higher degree of reliability. Thus the excess capacity can be lower, which means lower capital costs and lower operating costs. It is a kind of rethinking of the economies of scale. Furthermore, small plants are easier to find sites for: they emit fewer pollutants and can be built on sites that could not accommodate a larger plant. And since a smaller power plant takes a relatively shorter time to construct, the degree of uncertainty in long-range forecasting of demand is diminished, thus diminishing the potential of overbuilding capacity.

Much the same appears to be true with nuclear power plants. A study conducted by Clark University showed that nuclear plants of 800 megawatts or less "have distinctly better performance records than large plants. . . ." Atomic Energy of Canada, Ltd., a government-owned nuclear company, has recently announced plans to develop the "cheapest and smallest reactor ever designed for commercial use." It is called "safe low-power critical experiment" and nicknamed "Slowpoke." It will produce hot water to heat buildings, not steam to turn turbines, and it will, they believe, produce a thermal kilowatt for $425, which is well within the current range for the equivalent from a conventional nuclear reactor. The entire installation, it is believed, would cost less than a million dollars. It would not require expensive core-cooling safeguards and could be used anywhere that petroleum is expensive and district heating systems are practical.

In addition, Rolls-Royce, a company long associated with luxury automobiles, has decided to market small nuclear reactors. A management decision was made in 1979, following the Three Mile Island nuclear accident in Pennsylvania, not to proceed with company plans to begin European sales of 1,300-megawatt nuclear reactors built by the U.S. Combustion Engineering Company. Instead, the company's management scrapped plans for the large plants, and concentrated on building a nuclear plant one-fourth the size, that could be prefabricated and barge-mounted. According to Peter H. Jones, director of nuclear projects for Rolls-Royce Ltd., new company plans call for standardizing the size of their barge-mounted plant at 300 megawatts. This would make the Rolls

reactor one of the smallest designed for commercial operations since the 1950s. The standardization concept includes
not only the use of the small, prefabricated units, which can
be prelicensed by regulatory authorities, but also the use of
modern assembly line techniques developed in England's oil
platform construction industry.

The managers of the Rolls company were intrigued by
the accomplishments of billionaire Daniel K. Ludwig, who
installed a floating pulp mill and associated barge-mounted
power plant on his agricultural super-plantation in the heart
of Brazil's Amazonian rain forest. The new company strategy
calls for installing these small reactors in growing Third
World countries, but the reactors might also find markets in
many areas of the industrialized world.

Similar plans have been projected for fusion plants, suggesting that if we consider only large fusion plants, we will
lose the likelihood of competitive designs and concepts. C. P.
Ashworth, a mechanical engineer with California's Pacific
Gas and Electric Company, argues that with "the small facility focus, many inputs get into the act, including rivalry and
competition between institutions pursuing different projects
—the small facilities route can lead us to attractive commercial fusion energy sooner."

Long lead times, high capital cost, shrinking economies of
scale, problems of reliability—all of these matters are widely
recognized to be unavoidable albatrosses for big power
plants. That they cause a reliance on vulnerable fuel supplies
has also recently become better understood. That they present an adversary with ideal targets is not yet so well known.
It is time to find substitutes.

SUMMERTIME

Corporate advertisements in the early 1970s were likely to
caution enthusiasts that conservation meant a return to the
Stone Age—an outrageous exaggeration. But truculent con-
servationists, who felt that the slightest intrusion on a favored

wilderness meant the final end of all human values, also exaggerated. And, as we have noted, President Reagan, in his words and actions, has treated conservation as an outmoded approach. It may be ironic that those corporate advertisers, among the first to declare their support for Ronald Reagan, were also among the first to move with considerable alacrity to conserve energy in their back yards, even if decrying it elsewhere.

With the problems of gigantism and centralization increasingly obvious, with fossil fuels supplies inevitably on the decline—even without a catastrophic interruption—and with the demand for energy in this nation *inevitably increasing,* it behooves us to conserve and to eliminate waste. Not using energy—in certain ways—is in fact the best fuel source we have; it can result in increased energy use in those areas where it is needed.

We are inclined to think of energy as a product, and to an extent, of course, it is: a certain number of barrels of oil refined, a certain number of tons of coal mined, a certain number of kilowatts delivered. But as Roger Sant of the Mellon Institute has pointed out, producing barrels of oil or turning the thermostat up or down seasonally—production and conservation—"only partially addresses the function of energy in our economy and lives. A thriving economy and a materially rewarding life are dependent not on the given quantity of energy consumed, but on the *services* or *benefits* that are derived from that consumption." It is in this context that we should explore the matter, broadly speaking, of energy conservation.

Industry, which accounts for about 40 percent of American energy consumption has, in general, done the best job of conserving and increasing efficiency. In its U.S. refineries, for example, Exxon reduced energy use by 21 percent between 1972 and 1977—and the large bulk of these savings were accomplished with little or no capital investment. In the same period, Lockheed's Los Angeles factory complex reduced energy use by a whopping 59 percent, again with little or no capital expense.

Those, and other stories like them, are the good news. The bad news is that it will be harder from here on out. Thomas C. MacAvoy, president of the Corning Glass Works, was quoted in the *New York Times* in January, 1981, as saying, "We have done the major things that conserve energy. Now we have to do things that are capital intensive." Estimates of the capital costs for American industry to increase efficiency and remain competitive range from $40 to $80 billion—sums that could strain the U.S. capital market. As *Times* reporter Peter J. Schuyten pointed out, for capital-short industries such as steel it may simply not be possible to make the needed changes, whereas in others such as "the petroleum for petrochemical industries, where cash flow is not a problem and financings are common and easy to come by, the burden will be less taxing."

Some federal laws have provided further incentives, such as the 10 percent business energy investment tax credit established by the 1978 National Energy Act. However, the Internal Revenue Service, in what is considered "a major setback" for industry, recently issued proposed rules to determine which technologies qualify for the credit; the new rules did not expand the list.

It had been hoped in some circles that sufficient energy conservation would take place merely as a result of rising energy costs, but Robert Stobaugh and Daniel Yergin, writing in *Foreign Affairs,* suggest that "such a course will not be adequate. The gap between energy resources and energy demand would be closed by 'unproductive conservation'— the shutting down of factories, higher unemployment, higher inflation, offices too warm in the summer for efficient work, colder houses, a choice for some between food and fuel. . . . Far more desirable is the alternative of accelerated energy efficiency."

Much of the remainder of this book will be devoted to

analyzing just how to accelerate energy efficiency, which would have a variety of strategic effects beyond the obvious one of reducing our dependency on foreign resources. Energy efficiency could decrease pressure on centralized facilities and reduce the need for new highly capital-intensive construction of such facilities, thus freeing up capital for other areas of the economy; it could reduce the demand for strategic materials; and it could reduce inflation. There are, however, certain kinds of institutional barriers in the way of an intelligent and aggressive conservation policy. The hydra-headed approach of the federal government leads to confusion, to be sure, but consider the electric utility and its view of the world: since the utility passes along the cost of fuel to the consumer, a dollar burned is a dollar earned. What incentives are there for a utility to assiduously seek to be a no-growth industry?

Furthermore, conservation is not sexy; production is. Consider how easily Congress was persuaded to provide federal subsidies of $88 billion for synfuels in the 1980s. It is estimated that the government provides $20 billion a year in energy subsidies. One of the nation's foremost thinkers in the field of energy conservation, Dr. Arthur Rosenfeld of the Lawrence Berkeley Laboratory, calculates that half of these subsidies, redirected toward conservation over the next ten years, would cost the nation a total of $100 billion and would provide enough interest-free loans to insulate half of America's housing stock. This, Rosenfeld points out, is equal to the oil we now import that would be replaced by the synfuels industry. Perhaps the insulation manufacturers don't have a big enough lobby.

The psychology—and reality—of inflation breeds an odd sort of there's-no-tomorrow attitude in some people. Why invest my capital in something that may or may not ultimately pay off? Furthermore, I have no capital to invest. I'll just lower my standard of living and try to keep up. These thoughts followed a conversation we had with a young man named Don Walker. Unknown outside of Loudoun County, Virginia, he is a local building contractor with a special interest in alternative energy and conservation techniques. Among his specialties is the insulation of homes.

"You know who insulates their houses?" he asked us the day he stopped by to have a look at that heat pump in the basement. "I'll tell you who insulates their homes. It's the poor guy, the farm laborer. Or it's the rich guy. The people in the middle class, they're mortgaged up to their necks, they've got outstanding loans, and the food bill is going up like a rocket cause the kids are eating more now that they're bigger. They haven't got a cent left after each paycheck, and there's no more credit, and they simply can't insulate their homes. The poor do it themselves, and there aren't many rich people."

THE BOTTOM LINE

And yet there is a trend toward greater efficiency and energy savings, and it will almost certainly accelerate. Industry will have to invest the necessary capital, however painful it may be to find it, and the beleaguered yeomanry will almost certainly insulate their houses before too long anyway, because we all face what has been termed "energy gravity"—the unavoidable facts of entropy, which we mentioned in chapter 4. The basic problem is simply that it takes energy to produce new energy. Most cheap and available accessible fuel deposits in this country have already been exploited, and the energy to exploit the rest fully may be equal to the energy contained in them. What is significant, and vital to our future, is the *net* energy contained in our fuel resources—not the gross energy. Net energy is what is left after the energy used in processing, concentrating, and transporting energy to consumers is subtracted from the gross energy of the resources in the ground, or the sun.

Drilling deeper and deeper in the ground for oil requires

more and more energy. Think of the energy costs involved in building the trans-Alaska pipeline. The story for natural gas is the same. Although there is a good deal of oil and natural gas in the ground, the net energy—our share—is decreasing constantly. Energy conservation—by which we mean a major shift to efficiency—is the only route to survival. Quick fixes—like turning off the lights in an empty room—while helpful, are simply not sufficient. There are numerous helpful quick fixes: many industries, perceiving their obvious financial self-interest, have made them; many commercial and residential establishment have not. Retrofit alternatives —or palliatives—include ceiling and wall insulation, storm windows and doors, heat pumps, weatherstripping, caulking, day-night thermostats, and pilotless natural gas furnaces. More efficient appliances and machinery including refrigerators, water heaters, and other large energy consuming devices are available. Energy savings from such equipment will come mainly through better engineering and construction standards—prompted by regulation. The State of California has achieved remarkably quick energy savings by mandating efficiency standards for appliances, home-building, and construction. And the state keeps a wary eye on the federal government to see that it does not lower these standards.

It is hard to play the prediction game with new energy-conserving technologies. Some government regulations have undoubtedly had a fundamental impact on energy conservation, but in the area of "prescriptive" standards—laws that dictate specific technologies that can be used—the record is blemished. Standards, for example, that dictate details of houses that can qualify for government loans have often not worked in different climate zones. New standards for energy-efficient appliances may not take into consideration the possibility of new thermal storage techniques, which we shall describe later. One might find the country in a situation where new appliances are developed on a mass basis with more energy efficiency, but without the needed capability to store energy during peak load periods. Thus, some energy would be saved, but the opportunity to minimize construction of new power facilities and minimize the use of imported oil to run peaking turbines would be diminished. There are, moreover, new technologies that offer the possibility of substantially reducing imported oil use in homes and buildings—without need of extensive insulation retrofit. The Volkswagen Corporation has been quietly testing an oil-fueled heat pump in several hundred houses in Germany that offers such a potential savings. The VW heat pump uses the engine in the Rabbit diesel car to drive the compressor in the home heat pump system. Savings in the test are at least 50 percent over those from the use of a regular, high-efficiency oil furnace. And, since the car engine, rather than the electrical power plant, is running the compressor, excess heat is captured, allowing for a boost in the efficiency of the heat pump. This would be a boon for areas like New England, where electrical heat pumps do not operate efficiently in the bitter winters. The VW heat pump will operate in sub-zero conditions, because the heat given off by the engine supplements the heat pump cycle.

The most effective energy policy would encourage rapid turnover of obsolescent machinery and replacement of high-energy-consuming buildings and equipment. And a number of recent government and industrial analyses have proposed to do just that. Figure 9 shows a chart that was developed by the "Demand and Conservation Panel" of the National Academy of Sciences' Committee on Nuclear and Alternative Energy Systems (CONAES). It projects levels of energy growth in the United States over the next thirty years under various assumed scenarios. The dotted line indicates very rapid growth (3.5 percent per year), such as that experienced in the past. Scenarios D, E, and F assume that present policies toward energy continue, causing the growth in energy consumption to accelerate. Scenario F assumes further that new government subsidies are added to energy, increasing the

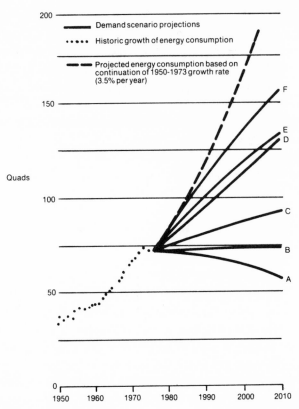

Figure 9. Demand and Conservation Panel Projections of Total Primary Energy Use to the year 2010 (Quads = Quadrillion BTUs)

growth in consumption. Scenario C assumes a slight rate of energy growth, triggered by moderate policies to increase conservation; the two remaining scenarios, A and B, illustrate the effects on energy demand that would be caused by increasing national policies toward conservation. Only Scenario A, assumes that aggressive policies and attitudes toward energy conservation would be accompanied by any lifestyle changes on the part of the public. In other words, simply adapting to higher prices by using newer, more efficient technologies offers the promise of substantial energy conservation by families in the United States.

Energy conservation is in fact a brand new field, and no one knows precisely how the nation should approach it. The federal government has invested precious little money in research in this area, and there are numerous puzzles to be solved. The National Academy of Sciences has estimated how long it would take to turn over energy-inefficient equipment and buildings and replace them with efficient ones: in a sense, to rebuild America along lines more desirable and necessary in an age of dwindling resources. They suggest the transformation would take more than 50 years in housing, 20–50 years in industrial plants, and 10 years in the U.S. automobile fleet.

THE SPECIAL CASE OF THE AUTOMOBILE

Transportation amounts to a quarter of the national energy budget; of this, more than half is accounted for by people driving around in cars. And anyone who thinks the American public is going to stop driving around in cars in the forseeable future simply does not understand the American public, or the obvious and basic facts of American demography, or even the CONAES estimate above. As long as the residential patterns of the United States are what they are, the automobile will continue to be essential. But it does not have to be the absurd machine that it currently is.

The major gains in American automobile efficiency have been the result of weight reduction and the importation of foreign technology—a sad commentary on the nation that spawned Yankee ingenuity. As Stobaugh and Yergin pointed out in *Foreign Affairs,* "substantial technological innovation, as opposed to styling, has not been a major priority for the industry or its suppliers. Massive capital investment is needed over a decade for the four U.S. automotive companies. . . ." It is interesting to note, parenthetically, that last year, in 1980, Japan handily outstripped the United States as the world's largest manufacturer of automobiles. There are a lot of reasons why, but you cannot deny that part of this phenomenon results from the fact that more people wanted Japanese cars than American cars.

The U.S. Department of Energy has accepted as the official energy conservation goals of the nation the most optimistic scenario among several considered by the Committee on Nuclear and Alternative Energy Systems. It called for an energy savings in the transportation sector of 38.2 percent in 1990—a decade away. The Chrysler Corporation, has been awarded a subsidy of over $1 billion so that it might stay in business.

Chrysler's government-subsidized marketing campaign in 1980 and 1981 skillfully argued the merits of the small, energy-efficient "K Car," a transportation miracle arriving just in time to help the U.S. oil import problem. What the television and newspaper ads did not mention was the fact that the U.S. auto industry had designs for these cars back in the 1960s but chose repeatedly not to put the cars into production. In *On a Clear Day You Can See General Motors,* John Z. Delorean, the General Motors' vice president who resigned abruptly in 1973, tells the story of GM's own K-Car, and why it was not produced. He notes that "All domestic car lines grew in weight and size during the 1960s. By the end of 1969, the combination of the surge of imported cars in the small-car market, our inability to generate substantial profit from our smaller car lines and the steady rise in production costs seemed to me to make this program necessary for the corporation. . . . The K-Car program proposed a common body and chassis for all the GM cars in the intermediate and compact car classes . . . [it] was aimed at taking weight and cost out of these car lines and improving their fuel mileage."

The K-Cars were to be designed more like European cars and would have been introduced to the public in 1973 or 1974 —coincidentally in time for the greatest demand for fuel-efficient transportation in the world's history. However, as

Delorean says, the program was finally shelved in late 1970, after considerable infighting among various GM divisions. "It never was formally turned down, just permanently shelved," he says.

All of our domestic car companies missed the boat. The standard line in the industry, "small cars mean small profits," prevailed over that of the innovators like Delorean; the result is the precipitous decline of the U.S. automotive industry. Knowing full well that it is too late to catch up with the Japanese and the Europeans, the industry has recently pressured Congress and various political leaders to stop small-car imports. In early March, 1981, a delegation of state governors met with President Reagan to beg for protectionist legislation. Six of them—Milliken, Dreyfus, Orr, Thompson, du Pont, and Thornburgh—explained their position in a *Washington Post* letter: "One of the arguments," they stated, "against a negotiated quota is the assertion that Washington has no obligation to protect the auto industry from, among other things, failure to build the cars Americans want. The fact is the industry did build the cars Americans wanted. The cars were larger and consumed more gasoline than those produced by other nations. This was, in large part, due to the federal energy policy of keeping the price of gasoline below world market levels." The grand finale to this line of reasoning was that "Since the government helped create a demand for larger cars by creating an artificial price level for gasoline, it has a clear obligation to help the industry adjust now that it has changed signals and is allowing gasoline prices to rise to their natural level."

What these governors failed to report to the citizenry was the fact that the automotive industry has been the prime lobby in Washington to maintain artifically low prices on gasoline, so that it could continue to profit from guzzlers. In addition, companies like General Motors historically have done everything in their power—including the illegal, anti-competitive purchases of dozens of mass-transit train companies—to prevent energy-efficient options on the American market. In the case of mass transit, General Motors bought up a number of urban rail lines in the 1930s and 40s, simply in order to scrap them—trains, tracks, and all. In conjunction with the oil industry, this company was able to destroy mass transit opportunities in many cities—including Los Angeles, the black hole of automotive fuel savings—and the only suffering inflicted by the government was a few thousand dollars in fines, after the fact.

Today, the Detroit giants remain twisting in the vortex of an ever-deepening whirlpool of no profits, while the American public sensibly buys VWs, Toyotas, and Datsuns. The authors recall the times back in the early sixties when people —mostly on the East Coast—who were driving around in those strangely shaped German cars that came to be called bugs and got twenty-five miles to the gallon were considered weirdos. The VW drivers of that era would honk at each other as they passed on the highway, a kind of bond among a small band of strangers who recognized how smart they were. The folks in Detroit should have paid attention.

Monday-morning quarterbacking is one thing. Choosing to field a team of known cripples is yet another. Certainly, a nation that had the flexibility to convert an auto industry to one that made thousands of tanks a year in World War II can find the will to produce a decent, appropriate automobile. If it cannot, others will do it for us.

One company with a sophisticated approach is Volkswagen. Already, the Volkswagen diesel Rabbit has the highest fuel economy of any car sold in the United States—a combined EPA rating of 50 miles per gallon. This means that the Rabbit uses about one-third as much fuel as the average car on the road in the United States. Add to that the fact that diesel fuel does not require as much energy in the refining process as does gasoline and you begin to see the dramatic implications of a major shift to cars getting this kind of mileage. Not content with 50 mpg, however, Volkswagen engineers like Gerhard Delf, the mastermind of the diesel program, have been working on a whole new generation of vehicles that promise future migraine headaches for the guzzler builders.

Volkswagen has been working on a new vehicle, based on the Rabbit diesel but using several improvements, that will get a *combined* fuel rating of 80 miles per gallon. The car, which looks no different than the ordinary Rabbit, weighs 2,000 pounds—about 300 pounds lighter than today's diesel Rabbit—and employs a direct fuel injection system on its three-cylinder diesel engine. The supercharged diesel engine is also connected to a special system that cuts off power when the accelerator pedal is not depressed. In tests conducted in conjunction with a German government research effort, the prototype car has gotten nearly 100 miles per gallon—on the highway. Volkswagen has not decided yet on a production strategy, but the car could be available on the U.S. market, and made in a U.S. factory (owned by Volkswagen of America), within three or four years. This achievement would—at one stroke—radically alter the structure of the U.S. transportation industry.

Japanese car-makers are not far behind the Germans. At least one company has developed a diesel auto similar to Volkswagen's prototype Rabbit, but members of a National Academy of Sciences' study team say that the Japanese executives told them there would be no exports of Japan's 100 mpg car to America. When asked why, the Japanese laughed and indicated their belief that introducing such an efficient vehicle to the United States would cause a trade war. Their judgment is not flawed.

Although the U.S. Congress and the automobile industry may be able to stave off the foreign imports temporarily, improving the fleet fuel economy remains the single most important means by which we can eliminate the OPEC oil threat within this decade. This means that making the ways that Americans travel from place to place more efficient is a

vital national security objective; it is certainly more important than the economic consequences of not limiting foreign
car imports. In fact, the wise Japanese and German car
manufacturers should be rewarded, not penalized, for helping the United States reduce its energy security problem.
Introducing a new era of transit efficiency to the United
States will come about largely as a result of increasing fuel
prices—but the U.S. consumer is now paying less than half
the price of retail gasoline in Europe and other industrialized
parts of the world. This is due to long-existing taxing policies
that have placed a heavy premium on gasoline. As a result,
more and better transit opportunities are available in
Europe.

A number of observers, including Marc Ross and Bob
Williams in *Our Energy: Regaining Control,* advocate a national shift to an energy tax, in order to spur conservation
efforts. This tax, they say, "would be accompanied by a coordinated rebate or reduction in other taxes, so that typical
people and firms would stand still, so to speak. It is important
to keep funds in the hands of energy users so that they will
be in a position to invest in energy-efficient improvements.
While a tax shift would generate no new revenues, it would
give the nation the opportunity to reduce those taxes which
are economically inefficient in their impact, onerous to ad-
minister, or highly uneven in their application. Sales taxes,
social security taxes, individual and corporate income taxes,
and property taxes are candidates for reduction in favor of an
energy tax." Ross and Williams would begin with the introduction of a gasoline tax, which would be gradually increased
to a level of $1.20 a gallon. Even this level of taxation, how-
ever, is a third less than existing gasoline taxes in France and
Italy. Certainly, such an approach offers the only viable, long-
range approach for the federal government to both increase
revenues for such purposes as mass transit and encourage a
shift to efficiency through tax policies.

As the cost of fuel rises, some efficiency improvements

will result in the transportation sector. Already, airlines are scurrying to purchase more fuel-efficient planes, like the new Boeing 767s and 757s. However, even the more efficient aircraft, which save about one-third of the energy used by conventional jets, may not prove to help much—as OPEC costs escalate, and as people can no longer afford air travel. A major area for savings would be the improvement of our national rail beds and the encouragement of shifts of freight transportation to rail. Moving freight by rail takes one-fourth the energy required to move it by trucks, yet our national policies encourage more highway freight movement and discourage the movement of goods by rail. We look enviously at the superior rail facilities of Europe and at Japan's famed bullet train, but no major commitments—either public or private—have been made to improving the U.S. rail transportation system.

International lawyer Frank Davidson, who specializes in giant engineering projects, tells the story of his attempt to arrange for private financing of a high-efficiency train program for the northeastern U.S. rail corridor. The financing for the project would have cost the U.S. government little, but the federal government opposed the project anyway, citing the potential for other options—hovercraft, planes, and so forth. This happened only a few years ago, but the tale indicates the lack of serious attention paid by our government to the need for energy-efficient mass transportation.

In *Energy: The Conservation Revolution,* Bill Chandler and Jack Gibbons point out that our nation could develop a fleet of vehicles averaging 40 miles per gallon within the next ten years at a cost of $60 billion. By halving automotive fuel consumption, over two-and-a-half million barrels of oil would be saved each day in the United States. "At the same time," they argue, "a fleet of automobiles would have been produced capable of burning a less expensive fuel, one that could be made from a variety of sources. If the U.S. government desires to fund any crash energy program, it should be one to increase the fuel efficiency of automobiles above 37 mpg as soon as possible." Such a relatively small cost invested in our security could harvest rich dividends. It should be the goal of national policies and industrial policies as well. Only through such a major national commitment can we equip our transportation industry to meet the needs of the future. A combination of policies, ranging from energy taxes to reevaluating the government's role in rail transportation, will be required, but at center stage will be the improved efficiency of our national fleet.

SMALL IS EFFICIENT

A number of national leaders in industry and government have recognized the need for increasing the efficiency of our economy and minimizing the extravagant waste of energy that is deepening our energy security problem. Republican Senator Malcolm Wallop of Wyoming is one of the leaders in this movement. Wallop says, "My interest in energy conservation stems in part from my committee involvement in both intelligence and energy areas. The national security risks created by energy dependence force us to look for the most immediate solutions to our energy problems. . . . Energy conservation, combined with increased exploration for oil and gas, provides the most immediate response to our national security problems. The point is, we must proceed on all fronts simultaneously. There is no single solution to our energy shortage, but increasing industrial energy efficiency must be recognized as a crucial component of our energy policy." To this end, Senator Wallop has sponsored legislation in the Senate to provide a 20 percent tax credit to industries for installing a variety of devices to cut energy waste. The bill is called "The Industrial Energy Security Tax Incentives Act of 1981" and is co-sponsored by a number of other politicians, including Democratic Senator Ted Kennedy. When Wallop tried to get such an approach through the Congress during

Jimmy Carter's administration, he was opposed by the president's lobbyists—who argued for conservation but failed to recognize the critical need for American industrial efficiency.

William Gould, president of one of the nation's largest utilities, the Southern California Edison Company (SCE) in Los Angeles, has endorsed the need for greater decentralization and reliance on smaller, more efficient power facilities. In 1980, Gould announced that SCE would abandon building conventional baseload power plants in the near term and would seek one-third of its generating capacity additions for the 1980s from "new and alternate" energy resources. Gould and a team of specialists found that the utility would have trouble financing and building massive electric plants and that "it was not feasible to base resource additions on facilities with high capital costs, long lead times, and long gestation periods before construction begins." SCE is committed to a diverse program, testing a variety of end-use conservation options, building cogeneration plants, fuel cells, and solar and wind plants, and extending the company's hydroelectric capacity.

What Southern California Edison is pioneering is a fundamentally new approach to the energy problem—using a number of diverse energy sources in new ways to increase efficiency and meet consumer's needs. In the following chapters, we will look at some of the technologies that use a myriad of smaller-scale solutions; they are the building blocks for a more efficient and less vulnerable society.

8. Industrial Strategies

MUCH OF the work to be done to render the United States less vulnerable and to improve the security of our energy-threatened economy will be difficult, uninspiring, and quite unlike building up massive dramatic efforts like the space program. A good deal of it will be like doing household chores.

DEMANDS

One of the most compelling areas for energy savings is called electrical load management, which is a specialized form of accounting. Load management refers to methods that can be used by utilities to reduce the peak needs for electricity within systems. Unlike the phone system, which is designed to handle most—but not all—telephone calls during peak periods (like noon on Christmas Day), the electric system has traditionally been designed to meet the peak demands of consumers. People turn on their irons, furnaces, air-conditioners, or whatever at certain times of the day, creating daily peaks during which the local utility has to provide its maximum amount of power. Businesses and commercial establishments also create peaks. Load management also relates to the fact that machines, in the nature of things, sometimes fail to work; utilities then have to build more generating capacity than their actuarial tables of peak use suggest is going to be used at any given time. This frequently means that inefficient power plants burning imported oil have to be used during those times when air-conditioners are busiest in summer and heating needs are at peak in winter.

Rather than use inefficient, older plants to meet peak electrical needs, some utilities are using what might be called "synthetic hydroelectric" power. During the times of day when people are using a relatively small amount of power, the utility will use the excess power available on the grid to pump water uphill, storing some of it in a special reservoir. Then, during peak periods, the water is released from these "pumped storage" facilities, and it passes through hydroelectric turbine generators. However, such solutions to the peak electricity dilemma have not found favor with environmentalists, and many projects have been delayed or blocked. There is also the familiar net energy dilemma: it takes a third more energy in the overall equation to pump all that water up into the reservoir, so the net implications of pumped storage are not attractive. There is a considerable energy penalty.

There are, however, a number of new and potentially dramatic ways to solve the peak problem; the answer lies in ways of *reducing* peak demands and making the electricity grid itself more efficient, so that electric loads can be managed on a twenty-four–hour cycle. Some utilities, like the New England Electrical System companies, have called for the implementation of new technologies to reduce peak demands. In this system's long-range planning document— NEESPLAN—the load management concept is defined as follows: "New England Electric is currently a winter peaking company [as opposed to utilities in the south with high summer air-conditioning loads], and New England is a winter peaking region. Therefore, reducing the winter peak through the use of storage heat is an important factor in our future plans. Storage electric heat is an economically viable and proven technology. Through the use of storage heat, essentially all of our new electric heat load can, if our programs are successful, be in off-peak periods. . . . Metering and communications systems are rapidly developing technologies. Further advances will be forthcoming, which should greatly reduce the cost of load management. An intensive conservation and load management program has the potential of lowering utility costs and increasing service reliability. Lower load growths and higher load factors will result in more efficient utilization of existing generating capacity and will reduce new capacity requirements." According to the company's plan, the peak growth will be held to 1.9 percent per year, and the overall program will reduce energy needs by an impressive 60,000 megawatt-hours of electricity per year.

Ways to accomplish this include a variety of new technologies, involving the use of sophisticated microelectronic devices to monitor and control electric loads and the use of new rate approaches to encourage consumers and industries to use electricity in off-peak periods. In addition, there will be a need to develop new ways to store energy—usually as heat—during short peak periods, to eliminate the need to operate power facilities.

One way to lower peak demand is by what are called "time-of-day" rates. In other words, at times of peak demand, you pay more for what you use. Such rates are directed chiefly at residential consumers of electricity to encourage them to reduce their use of energy-intensive appliances such as hot water heater or air-conditioners. Time-of-day rates were encouraged by the Public Utility Regulatory Policies Act (PURPA) of 1978, as were seasonal rates and other techniques, in the hope of getting utility rate structures to reflect the actual costs of providing electric service. Electric water heaters are prime targets for peak reduction practices; it seems that most people take a shower and wash the dishes at the same time of day, creating a "high-coincident-demand."

Translated into power-plant terms, the electric water

heater equals about 4.5 kilowatts of inferred capacity. If the utility were to build new power plants to meet the peak demand caused by water heaters the cost per household for installed capacity, not fuel, would be over $1000. Using load management techniques and thermal storage techniques—which will be discussed soon—the utility could save the capital cost of building a new power plant; this conservation alternative would cost it only $200 per household. Translated into terms of thousands of households, the savings are potentially enormous. Who would control the use of the energy inside the house? Say you put a timer on the water heater to see to it that it runs only in off-peak times of day. Would you control the timer or does the utility?

Two officials of the Wisconsin Public Service Commission opt for household control, coupled with time-of-day rates. The householder, they say, "would not try to cheat himself, since the conservation strategy would reside in his imagination, not in some distant utility boardroom." On the other hand, one of the problems with an overall load management strategy for a region is that we do not really know how people use their electricity. About 90 million electric customers have their meters read once a month; those readings comprise an enormous data base, but it does not really tell the utilities how the electricity is being used. Remote monitoring, using computer microprocessor-coupled communications systems, could provide such information, which would be invaluable for load management plans.

The Southern California Edison Company, a major private utility, is currently pursuing a program called Demand Subscription Service whereby a demand-limiting device is connected to the utility meter at the residence. When the household's demand for power exceeds a prearranged amount, electrical service is automatically disconnected. The householder, no doubt chagrined, locates whatever appliance is the culprit, turns it off, and then manually switches on the juice for the rest of the appliances. By the end of 1981 there will be some two thousand of these devices installed in southern California. SCE has also established a new energy cooperative concept for larger commercial customers. The first of these was formed in 1979 in Orange County. Several large companies and a shopping plaza formed the Orange County Energy Cooperative Association. For a monthly rebate of $120,000, the co-op agrees to shave four megawatts off the peak load whenever the utility requests it to do so. The co-op receives a million and a half dollars in annual rebates, and the utility saves $4 million on the initial capital investment.

Another approach to load management is being tested with residential customers in Ohio. Buckeye Power, Inc., a public utility, has instituted a comprehensive program of electric water heater controls with radio switches that curtail use during peaks. Since 1975, when Buckeye started the load management program, about 20 percent of their customers have taken advantage of the program, totalling over forty-five thousand homes with the water heater controls. Less than half of the individual electric cooperatives who get electricity through Buckeye use financial incentives to encourage customers to participate, but the incentives are only in the $1 to $2 per month range. Most of the co-ops simply tell their member customers that they can hold down wholesale costs of power by adopting load controls. In 1980, Buckeye saved $2.5 million in purchase power costs (from private utilities) by using load controls to reduce the peak demands in its system.

In addition to employing load management strategies, however, it will be necessary to store energy in a variety of ways, for release during peak periods.

EXCESS ENERGY

There are three basic ways to store energy: as heat, as electricity, or as kinetic energy. Some suggestions for the efficient storage of energy follow below.

The generation of electricity produces hot steam, which is, more often than not, dissipated into the environment. But in many instances it could be used for district heating. Moreover, the heat generated by producing electricity in summer could be stored and added to the winter heating capacity.

A lot of heat is already stored up for our use—in the water in natural underground aquifers. Geothermal radiation usually regulates an aquifer's temperature: in winter the aquifer is warmer than the air and in summer it is cooler. Heat pumps or heat exchangers can exploit this convenient temperature differential. Experimental heat pumps in aquifers run with phenomenal efficiency, and further studies are being made to determine the potential performance of these enormous storage tanks. Even manmade aquifers are being tested.

Energy can also be stored chemically in an electrical system—in batteries. The conventional lead-acid battery cannot withstand the constant cycling between the fully charged state and the discharged state that is essential in either utility or electric auto use. To produce a heavy-duty design would be prohibitively expensive. Other batteries designed for high-temperature use—such as lithium-sulfur and sodium-sulfur cells—are receiving a good deal of attention, but they have serious technical problems associated with their design, and materials and may well not become commercially useful. On the other hand, the Department of Energy and NASA are experimenting with a battery system called Redox, for reduction-oxidation; it promises to be an inexpensive, long-term, and reliable method of storing electricity. It is a very simple system consisting of a "stack" or combination of cells that takes advantage of the change in valence in the reduction-oxidation process. The Redox battery uses chromium and iron chloride as reactant fluids and inexpensive carbon electrodes. In addition, it can operate at room temperature. Currently, the Redox is being developed for use in the kilowatt range, but these systems could be scaled up for use in utility load leveling. In the meantime, a variety of other novel battery technologies are being tested for load management applications.

Yet another way to store electricity is in magnets. In a typical electromagnet, the resistance of the magnet's windings causes power losses; power, therefore, must be constantly applied in order for the magnet to retain its field. But if the windings lack resistance to electricity—if, that is, they are made superconducting—then once the desired magnetic field is created, no further energy input is required. The original energy input is thus stored in the magnet with up to 95 percent of it available to be drawn off as needed. Studies at the University of Wisconsin and at Los Alamos Scientific Laboratory suggest that such a method would be economical only for utilities operating in the 1,000 to 10,000 megawatt range.

Recent developments in materials and design have made it possible to use sophisticated flywheels for storing energy in utilities and in vehicles. Huge flywheels, spinning in an inert gas to reduce friction, could store as much as 20,000 kilowatt hours of energy. Small ones in vehicles offer the added advantage of being able to be recharged far more quickly than batteries; the brakes would become a regenerative device for the flywheel, extending an electric car's range by 25 percent.

Hydrogen presents itself as a storage medium. Using the excess energy from off-peak periods, water could be electrolyzed into hydrogen and oxygen. The hydrogen can be stored—albeit at high costs and with the use of exotic materials—and, during periods of high demand, used to power fuel cells. Also, the use of ordinary compressed air represents an attractive storage opportunity. In a conventional gas turbine, compressed air is mixed with the fuel to generate mechanical power—in this case, to turn the turbine blades. About 60 percent of the energy produced by the turbine is needed to operate the air compressor. But during the off-peak periods, compressed air could be produced and used later to operate the turbine.

Figure 10 shows the costs of three alternative utility energy storage technologies: advanced batteries, underground pumped storage, and compressed-air storage. All of these techniques look economically promising for periods of energy discharge (during peaks) of up to eight hours duration. Utilities can purchase storage systems that use these three techniques for costs under $500 per kilowatt of new plant capacity. This is a considerably lower investment than what would be necessary to build new power plants to meet peak needs.

EUROPEAN EXPERTISE

The world's leaders in load management and energy storage are the Europeans, who have implemented sophisticated systems ranging from Britain to Germany. After the Second World War, the Germans took advantage of air-raid warning systems that had been used to turn off streetlights in the cities, when Allied bombing raids were imminent. Rebuilding their war-torn economy was particularly difficult, since their energy systems had been severely crippled by the 1944

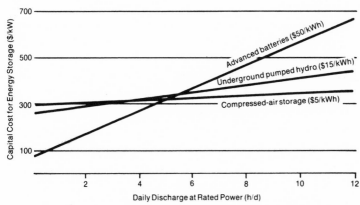

Figure 10. Comparison of Energy Storage Technologies

bombing. However, the civil defense systems provided a convenient way to make maximum use of the electric power grid—the controlling devices were used to shift electric loads and to control various kinds of heating systems.

Figure 11 shows how one municipal North German utility, the city of Hamburg, managed to use a combination of electric storage heaters and load management controls (to use electricity at night and release heat during the day) to essentially flatten the daily winter peak load curve. The chart follows a typical January day from early in 1968 to the later part of 1973. Unlike the utility's American counterparts, who have simply built more power plants to meet peak load demands, Hamburg is a model of energy efficiency and wise control of the electrical grid. According to a recent article in *Technology Review* by Argonne National Laboratory researchers J. G. Asbury, R. F. Geise, and R. O. Mueller, the introduction of special electric storage heating technologies in Europe has been very rapid, even though the units may

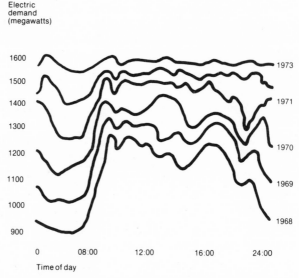

Figure 11. German Load Management: 1968–1973

cost $1000 more to install in a home than simple electric
resistance heating. This was commercialized, they say, in
Britain and Germany by the introduction of special off-peak
rates for separately metered heating systems. "In return for
the low rates, the utilities obtained control over the charging
cycle of the storage devices. In Britain, utilities have generally favored switching by local clock control. In Germany,
they have used pulse-coded signals injected into the supply
network and picked up by receivers on the customer's heating system. Under either mode of control, the utility is constrained to operate the charging cycle in order to assure
customers the required amounts of thermal energy for space
heating. In both Britain and Germany, market penetration
was very rapid once off-peak tariff incentives were introduced. In each market region, installed capacity rose from
less than 800 megawatts of electricity in 1963 to approximately 20,000 megawatts of electricity by 1973."

ELECTRONIC ENERGY MANAGEMENT

Today, the explosive growth of companies in the energy and
load management field in the United States portends a similar future for this country. Unlike the European system,
which has operated well for decades through conventional
technologies, the American approach to the problem is
rooted in the electronics revolution. A 1981 survey conducted
by *Energy User News* counted dozens of firms, ranging from
small, local tinkerers and inventors to industry giants like
IBM and AT&T. According to the survey, "increased user
interest in wireless energy management systems is expected
to contribute substantially to the strength of the market,
which, manufacturers predict, will grow 11 to 25 percent this
year. The introduction of smaller systems, growing user
confidence in computerized equipment, and rising oil prices
following decontrol by the new administration are also expected to boost sales."

These systems are used in a myriad of applications. *Energy User News* gave a typical example: "At the National Bank in Lakewood, Colorado, a wireless system installed last June [1980] has been providing a 32 percent return on investment, according to bank vice president, Barry Willett. The $5400 system, manufactured by Functional Devices, Inc., cycles HVAC (heating, ventilating and air-conditioning) loads in four zones in the two-story, 10,000 square-foot bank. . . . The system has been trimming the bank's monthly utility bills by $300 to $350 in summer, and by more than $400 in winter, said Willett." Systems like this one control heating and cooling equipment by sophisticated tiny microprocessors that store information and time the operation of the equipment. The electronic instructions are sent via the building's existing electrical wires to the energy machinery, eliminating the cost of rewiring an existing building for energy controls.

There are other ways to transmit signals, including existing telephone lines, and the Bell System, prompted by deregulation, is looking forward to a huge new market. Bell is marketing a new concept called Energy Communications Service (ECS), which is sold under the trademark *Dimension PBX.* In their words, "ECS solves a different communications problem for you. ECS speaks to your energy-consuming equipment—to control energy usage in the right amounts, at the right times, in the right places. The result is a highly sophisticated system designed to help you use the least energy at the least cost. . . . In fact, it has been shown that ECS can help reduce annual fuel and utility bills up to 20 percent or more. In some energy retrofits, this cost saving has given customers a payback in less than one year." Bell is counting on a major marketing factor, the existence of the phone lines, to cut costs. "The cost of ECS can be only a small fraction of the cost of alternative energy management systems," Bell says, adding that "it can produce savings you never before thought possible."

Not to be outdone, IBM advertises its own energy management system, which is not only computerized, but in
buildings with computers, such as their own headquarters
building, the "energy management equipment recycles heat
generated by the computer installation itself. The outside
temperature has to drop below 11° F. before the boilers need
to be turned on." There is plenty of room in the market for
the little guys as well. A recent ad by the National Energy
Corporation of Burlington, Vermont, champions its energy
management system, *Solution 100,* with the following cap-
tivating offer: "Over 3,500,000 buildings are already Wired,
Ready and Waiting for a National Energy *Solution 100.* "

One company that has taken advantage of the new market for energy and load management is the Datrix Corporation of Larchmont, New York. Datrix has devised a technology using microprocessors and radio transmission of signals
between the power utility and the consumer to facilitate
"real time" control of energy-consuming appliances, like
water heaters, air-conditioners, and refrigerators. According
to Datrix president Brian Travis, Datrix now has "the technology to provide consumers with appliances that can readily
be adapted to sophisticated load controls. This offers the potential for substantial savings in electricity costs to homeowners and the utilities. The utility industry would prefer to have
dishwashers and electric water heaters that only operate during the utility's off-peak demand periods. Refrigerators and
air-conditioners can be equipped with thermal storage compartments, so that they can be operated during off-peak periods."

It will certainly be important for the United States to
develop new technologies for utility load management, since
these technologies can make a fundamental difference in the
ability of utilities to supply power in the future to consumers.
A recent analysis of America's electrical energy future by Dr.
Alvin Kaufman of the Library of Congress suggests that, by
1990, there may not be "sufficient capacity to assure the reli-

ability of the system." Dr. Kaufman notes that only 38 percent of the U.S. utility grid is centrally connected and that
power demands in some areas could exceed the ability of
local utilities to meet them in the future. Without central
connections, utilities in other parts of the country could not
transmit additional power to meet peak needs.

"As a consequence of the estimated high peak growth
rates coupled with inadequate construction expectations and
minimal current excess capacity, the faster growing areas are
projected to have inadequate reserve margins in 1990," he
says. The unique advantage of load management systems
combined with new technologies for local energy storage is
that this approach can be used in the immediate future to
make utility systems more efficient and to prevent blackouts.
Within a decade, if electric utilities and energy consumers
maximize the use of new technologies, they can have a pro-
found impact on electrical energy use in the United States.

Potential targets for load management strategies in U.S.
residences alone represent the tantalizing potential of up-
wards of 50,000 megawatts of electrical capacity. This means
that load management strategies to control domestic appliances and to equip them with thermal storage devices could
displace the power of fifty new 1,000 megawatt electrical
plants. Entirely new businesses may be created, and appliance manufacturers called on to equip air-conditioners, electric heaters, and water heaters with new technologies to
meet growing needs.

FIGHTING ENTROPY: COGENERATION

The idea could not be more straightforward. In a conventional system for producing electricity, steam is generated by
burning fuel to heat up water in a high-pressure boiler. The
steam drives a turbine that runs a generator that makes electricity. Along the way, a lot of steam is lost from the turbine.
In industry, some plants just produce steam from fuel. If you

built a system that did both—made electricity and processed steam—it would take less oil to produce the same amount of electricity and steam as the two processes separately: a savings of 15 percent. This is called cogeneration and there are plenty of ways of doing it, as is shown in figure 12.

One system, using what is called a back-pressure turbine, converts about 10 or 15 percent of its fuel to electricity—a relatively small amount compared to the amount of steam it produces. Adding a gas turbine to the system raises the amount of electricity four to six times—an efficiency that compares favorably to central power plants as electrical generators. Add to this the value of the process steam. Variations are countless. In one, a gas turbine produces electricity; a steam generator produces steam from the waste heat in the gas turbine exhaust; and a steam turbine uses the steam to produce additional electricity.

Industrial processes and machinery create large amounts of waste heat that is usually dispensed directly into the environment or processed by expensive cooling equipment. Yet the demand for waste heat is phenomenal. Such industries as food processing, textiles, pulp and paper, chemicals, automobiles, and some forty others all need it. In fact, these industries together use nearly 574 billion BTU of waste heat a year —about 20 percent of their total energy requirements. But waste heat can be extracted from other industrial processes, such as cement kilns, blast furnaces, and glass manufacturing, and passed along to its users.

Another development toward cogeneration is an energy device designed by the Fiat Auto Corporation of Italy; it ingeniously utilizes a small automotive engine to produce electricity in substantial amounts as well as heat. The device is called TOTEM—an acronym for Total Energy Module; it consists of a four-cylinder Fiat engine hooked up to an electrical generator and surrounded by equipment to effectively capture the waste heat given off by the engine. Fiat has sold scores of these engines in Europe, and the senior author has

A Conventional electrical-generating system requires the equivalent of 1 barrel of oil to produce 600 kWh electricity.

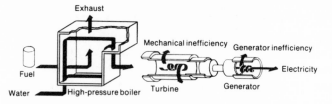

B Conventional process-steam system requires the equivalent of 2¼ barrels of oil to produce 8,500 lbs of process steam.

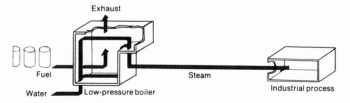

C Cogeneration system requires the equivalent of 2¾ barrels of oil to generate the same amount of energy as systems A and B.

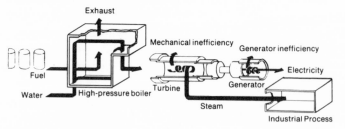

Figure 12. Cogeneration Processes

visited installations in Scandinavia, Austria, Italy, and the mideast. According to the TOTEM project's assistant director, Eduardo Bassignana, "The biggest problem in Europe has been the reluctance of the central power companies to

allow [Fiat's] small units in the power grid. One utility manager even told us that the power 'couldn't run backwards.' "
Yet the TOTEM energy converter has been successful, and the Brooklyn Union Gas Company has tested several in New York. Again, the only major problem was the opposition of the Consolidated Edison Company of New York to allowing installation of these sophisticated "micro-cogeneration" units in the grid. The State of California is planning to use TOTEM systems in a new research facility, in which air-conditioning units will also be attached to the basic system, allowing for greater efficiency.

Producing 15 kilowatts of electricity, the TOTEM converter's capacity is about four times the needs of a typical isolated household, but it could be integrated into multiple residential dwellings or used as a neighborhood resource. The waste heat energy could be used to heat the buildings. The TOTEM system has obvious applications in industry and agriculture. It is not only efficient and flexible but also cost-effective. It will provide electricity at an installed cost of $500 to $600 per kilowatt, as opposed to large power plants that cost up to $2,000 per kilowatt of capacity.

Meanwhile, the Thermo Electron Corporation of Massachusetts has proposed a similar system using mass-produced Chevrolet V-8 engines. It would be nearly as efficient as TOTEM but its advantage is that, using a larger auto engine, it can run at lower speeds, which results in fewer service problems. Also, the engine costs only about $15 per kilowatt. If it breaks down, it is economical simply to throw it away and replace it with a new one.

Cogeneration is not new. In Baton Rouge, Louisiana, the Gulf States Utilities Company, which is located in a petrochemical complex, has been producing both electric power and steam for the Exxon and Ethyl corporations since 1929. Several large petrochemical complexes in Texas are using sophisticated cogeneration systems. A General Foods plant in Massachusetts uses a bottoming cycle system: oil-fired boilers produce steam, which feeds a steam turbine generator that makes electricity. The low-pressure exhaust steam supplies a turbine generator that makes electricity; in turn, the low-pressure exhaust steam produced from that operation is then used in manufacturing gelatin and chemicals.

The Department of Energy estimated in 1978 that cogeneration could produce as much as 6 quadrillion BTUs per year in energy as a by-product of the U.S. industrial sector. Several other studies by the government and by industries such as the Dow Chemical Company have estimated that cogeneration projects can contribute up to 80,000 megawatts of electrical capacity to our economy—more than the current nuclear plant capacity. Such efficient fuel use would not only save dwindling fuel supplies and capital, it would also reduce both thermal and air pollution associated with electricity generation. Diesel generators coupled with various bottoming cycles are generally relatively small; they can, therefore, be placed near the site where the power is needed. The inefficiencies and unfavorable environmental effects of transmission are thus reduced.

While the environmental forecast for cogeneration is good, the economic one is mixed. The Department of Energy has stated that electricity from cogeneration compares favorably in cost with purchased electricity. But industry replies that commercial companies use a different set of criteria than do utilities for investing capital to produce electricity. For industry, the generation of electricity is essentially a sideline; in order for cogeneration to reach its full potential in that sector, a big shove is needed from the federal government (for example, Senator Wallop's additional 20 percent tax credit) and accelerated depreciation. With these incentives, paybacks can be achieved over a few years, which would encourage a wide range of businesses to invest in the new systems.

Another constraint on the immediate use of cogeneration by industry is the cost of standby power. If a cogenerator needs to supplement its own generation capacity, the utilities charge a high standby rate, reducing the project's competitiveness as a capital investment. Indeed, utilities have generally regarded industrial cogenerators as potential competitors or, at best, as energy liabilities they must have the capacity to serve. But the utilities, finding it hard to raise capital and to site new plants, are beginning to see that the cogenerators could be potential sources of energy for their own systems. Lower standby rates and credits are being negotiated here and there. In addition, the Public Utility Regulatory Policies Act has provisions that encourage the utilities to use and develop cogeneration.

The Federal Energy Regulatory Commission has proposed removing the restrictions against burning oil and gas in new cogenerator facilities—a move that would make such projects more economic. Other regulatory issues must also be resolved; for example, do steam and electric sales fall under federal, state, or joint regulation? Potential cogenerators are leery of the government's regulatory requirements on such matters as the issuance of securities for exchange. They also fear that projects with several industrial partners might run afoul of the Public Utility Holding Act of 1935, which was designed to control the abuses that were believed to be implicit in holding companies. This and other antitrust legal tangles are all slowing down the deployment of cogeneration equipment.

The U.S. government should be stimulating the development of cogeneration. Not only would these plants save primary fuels at a phenomenal rate, they could be mass produced in modular units, thereby enjoying distinctive economies of scale. Microsystems, like TOTEM, have the advantage of a pre-engineered design that could be mass-produced to suit a variety of end uses. They can be moved around from one location to another, operated on a variety of fuels, and used for emergency purposes. Ultimately, they have enormous strategic value for the United States.

A DIFFERENT KIND OF BATTERY

In 1839 an English scientist, Sir William Grove, placed tubes of hydrogen and oxygen in a sulfuric acid solution and joined them with platinum strips. The result was an electric current —and the principle of the fuel cell. In the early 1960s the National Aeronautics and Space Administration, looking for a highly efficient and reliable electrical generator of high energy density, applied the fuel cell principle to powering spacecraft. Soon fifty companies in the United States were following NASA's example; after several years, however, research showed that the first breakthroughs would not be sufficient to sustain commercialization and almost all the companies dropped the idea. Recently the U.S. Department of Energy has become interested in fuel cell technology, hoping to commercialize fuel cell power plants "for dispersed applications in the near-term."

Fuel cells are made of positive and negative lead plates and acid; they cause a chemical reaction that makes electricity as electrons flow through the acid (the electrolyte) from one plate to another. The problem is that the reaction causes all of these components to change from one chemical make-up to another—hence the need for recharging, by which you can restore the battery's components to their original state. Nor is recharging always an option in any given battery. In a fuel cell there are components that never change from one compound to another. With hydrogen in one side of the cell and oxygen (air) in the other, the plates and the acid react in such a way that an electric current is generated; the cells, however, do not change—provided that they have a continual fuel supply.

Naturally there are many complexities and a few disadvantages to fuel cells along with the advantages. But there are now a variety of electrolytes and other materials that can improve the efficiency and economy of fuel cells. As a low temperature technology, fuel cell emissions of particulates and sulfur and nitrogen oxides are far below the strictest governmental standard. They require no water for cooling; in fact, they produce water. They are highly dispersible and need fewer transmission lines because they can be assembled in modules, allowing them to be sited near the points of need. It would take only two years to build a fuel cell power plant, and while the cells currently operate with naptha or natural gas (from which hydrogen is produced in a fuel processor), a variety of other, renewable fuels such as biogas can be used.

The major limitation of fuel cells is that they rely on noble metals (usually platinum) for their electrolyte catalyst; platinum is not only scarce, it has competing uses, such as in the catalytic converters that reduce pollution from some automobiles. Nevertheless, it appears, according to the Electric Power Research Institute, that fuel cell technology will be commercially feasible by the mid 1980s and that an even more effective second generation technology will be feasible sometime after 1990. The chief advantage of this technology is that transmission considerations will make home use of fuel cells more efficient than fuel cell power plants—a siting option that few technologies can offer.

According to preliminary indications fuel cells can be produced more economically than central power plants and built in small systems sited in communities. Since they have no waste products—due to the chemical conversion of energy—they are ideal candidates for neighborhood power plants. In addition, they can be operated on a variety of fuels: ranging from coal-derived methanol to natural gas or even hydrogen produced by wind generators. They are virtually noiseless in operation; someday we may even see them in suburban basements. Utilities like Southern California Edison already anticipate that fuel cells will make a substantial dent in the energy needs of the 1980s and 1990s.

BUILDING THE NEW SYSTEM

The technologies discussed in this chapter represent both the old and the new. Intelligent use of them can contribute to our energy system the restoration of a badly needed balance. We cannot continue to build new power plants merely to meet peak needs of a few industries or extravagant consumers. Why not try the competitive approach and use some of the high technology that industry has spent years developing? The European experience in load management clearly shows a path toward a more efficient electrical grid, and the development of industrial techniques to cogenerate heat and electricity is a positive, and badly needed energy development for a society that needs new approaches.

Much of the problem in implementing the new technologies is an economic one. But "real cost" rates offered by utilities to consumers, based on deferring electrical loads to off-peak periods, would be a sufficient economic incentive to encourage high-technology industries to produce new methods for energy storage and load control. A growth industry in energy management is developing with tremendous speed; the same could happen with load management industries. A fundamental reappraisal of the structure of the utilities must be made by the utilities themselves and by the federal and state agencies responsible for their regulation. The waste of today's uncoordinated energy system can be turned into the profitable beginnings of a new system, integrating efficiency with economic growth.

9. The Uses of Solar Energy

DESIGNING BUILDINGS to maximize the use of solar energy is certainly not a new idea. The ancient Greeks took solar energy quite seriously and designed whole cities to take advantage of it. By the fourth century B.C., when Athens banned the use of charcoal—because the resultant air pollution was decimating the olive trees—solar energy was well on its way to determining the course of Grecian architecture and allied institutions. In one of the earliest recorded observations on solar architecture, Socrates, according to the historian Xenophon, stated that "In houses with a south aspect, the sun's rays penetrate into the porticoes in winter, but in summer the path of the sun is right over our heads and above the roof, so that there is shade. . . . we should build the south side loftier, to get the winter sun, the north side lower to keep out the cold winds."

The Romans rigorously applied Greek solar design principles—and improved upon them. In his ten-volume *De Architectura,* still a fascinating guide to applied solar concepts, the Roman architect Vitruvius laid out careful guidelines for the siting and orienting of houses, public buildings, temples, and even whole cities with respect to the sun. He stressed that the situation and construction of the individual structures, as well as the plan of the whole community, should fit both topography and climate. By the first century A.D., window glass had been developed in Rome, and this innovation made possible much more effective solar design and construction. Pliny the Younger described the *heliocaminus,* "solar furnace"—a room in his villa that had southwest windows, glazed either with mica or a crude form of glass. The idea of such a splendid heat collector was applied to agriculture as well, and the first greenhouses were built. Romans also applied solar design principles at the public baths— places of great social importance. The hottest bath, the *caldarium,* faced the south, and enormous windows covered the entire south wall of a typical bath house. Through careful design and inspired use of building materials, the Romans learned to trap solar heat in the floors and walls so that it could provide additional warmth in the evening. The Romans also developed the first legal concepts and laws protecting the access of buildings to sunlight, especially for their *heliocamini.* And to this day, certain rights to unobstructed sunshine are recognized by British law and by the legal codes of some other lands.

In North America, between 900 and 100 A.D., the early Indian cliff dwellers used solar design principles and techniques mastered centuries before by the Greeks and Romans. In our society, however, the application of sound solar design principles has largely been abandoned and replaced by the cheap energy of the fossil fuel era. Window air-conditioners, compact oil-fired heating systems, and other cheap technologies heating systems, have transformed architectural design. In recent years, as the heating and cooling costs of buildings have risen, homeowners, architects, and builders have reevaluated the older architectural concepts, including passive solar design.

PASSIVE SOLAR DESIGN AND ACTIVE SOLAR SYSTEMS

Solar energy systems for heating and cooling buildings and homes are either *passive* (mechanical devices and collectors are not employed) or *active* (pumps, collectors, and storage devices trap solar heat). *Hybrid* systems usually combine passive design approaches with a mechanical distribution technology. Solar space heating uses glass to trap heat, in conjunction with conservation techniques such as insulating shutters inside or outside the glass. Heat retention within a structure can be increased by the use of "thermal mass" or materials with high sensible heat capacity—such as rocks, water, tile, masonry, adobe, or materials that store latent heat, such as eutectic salts. Thermal mass heats up and cools off more slowly than air; its presence moderates the air temperature changes in the structure by absorbing excess heat during the day and gradually releasing it when the air temperature drops below that of the thermal mass.

In the passive approach to solar heating and cooling, the size and configuration of standard architectural elements are modified so they significantly contribute to the collection, distribution, and storage of solar energy in cool weather and the rejection and ventilation of heat in warm weather.

The simplest passive solar design is based on a principle of direct gain. The sun directly enters the living space through large double-paned, southern-exposure windows or through rooftop clerestories. The roof angle and overhang are designed to maximize entry of winter sun at a low angle and to minimize entry of summer sun at a high angle. The entering sunlight directly hits the storage materials, such as walls and floors, and transfers its energy to them. Overall, direct gain design lends itself to successful operation in areas with cool but relatively clear winters and with hot-dry summers. Cloudy days usually require back-up heat. Increased mass can offer longer storage, but in very cloudy and foggy climates it is not advised; the mass takes longer to heat and adequate storage may rarely be achieved—resulting in underheating.

Thus southern-exposure glazing is recommended to maximize winter solar gain, while external shading devices and vegetation minimize solar gain during summer months.

Floors and walls can be designed to store excess collected heat for nighttime or cloudy-day use in the winter. During the summer, the mass effectively reduces the daytime air temperature because of its ability to retain; it can then be "flushed" of heat at night through ventilation and re-radiation to the surrounding area.

A structure can be warmed or cooled by its relationship to local topography, sun angles, trees and other plants, ground water, precipitation patterns, and other aspects of local climate and geography. Such passive designs are highly cost-effective and can be used in a variety of applications, ranging from single-family homes to large commercial buildings. In New England, passively heated homes can easily provide 50–60 percent of the annual heating needs. In the Southwest, passive designs can provide virtually all of the heating needs in winter and, through the proper shading techniques, substantially reduce the need for mechanical cooling in summer. In addition, passive designs do not look like mechanical boxes rolled off an assembly line. Bruce Anderson of *Solar Age* magazine says, "the variety of climates and personal tastes will dictate a veritable plethora of unique solar home designs."

Active solar space heating systems, on the other hand, use glass-covered flat plate collectors to heat air or liquids that are circulated to a heat exchanger directly to the point of use or indirectly via rockbed, water, or chemical salt storage. Sunlight enters the collector usually through a glass or plastic glazing and heats the absorber plate—a black metal or plastic surface that is in direct contact with channels through which air or liquid is circulating. The collector is designed to maximize heat flow from a hot absorber plate into the circulating air or liquid. To improve performance and minimize heat loss from the absorber plate, the collector is covered with a transparent glazing, which reduces convective and radiant heat losses, and the absorber is surrounded with an insulated box. Coating the absorber plate with a special surface to reduce heat loss is also effective.

Flat plate collectors are generally mounted on a building or on the ground in a fixed position at prescribed angles that vary with the geographic location, collector type, and projected use of the collected heat. The optimum collector orientation for space heating, or combined space and domestic water heating, is due south. Ideally, collectors should be tilted up from the horizontal at an angle equal to the site's latitude for water heating. If the angles differ somewhat from this, the system will still function, but may require a larger collector area.

Liquid-type solar collectors commonly use water as the heat-transfer medium, with antifreeze and corrosion inhibitors commonly used as additives. The treated water carries heat from the collectors to an insulated storage tank. When heat is needed in the structure, it flows from storage through radiators, or air ducts, if a heat exchanger is used.

Air, as opposed to water, collectors are used in conjunction with rockbeds, water, or chemical salt storage. Warm air from the collectors flows either directly into the building's air circulation system or indirectly into storage. When the rooms are sufficiently warm or when the building is unoccupied, the heated air can be diverted to the storage bed, where more heat is stored with each pass through the collector. During the night or on cloudy days, heat can be removed from storage by circulating cool room air through a warm rockbed. Air systems are relatively easy to integrate with the conventional forced air heating system found in most homes and they eliminate the freezing, damaging, leaks and corrosion that can occur with liquid systems.

In these systems, back-up heating is needed when storage temperature drops below room temperature. The back-up may be a gas, oil, or electric hot water boiler, forced air furnace, or an electric heat pump. The back-up system has to be sized large enough to carry the total heating load during extended periods of cold cloudy weather. This can become a problem for electric utilities, which have to supply the make-up energy when the solar systems are not working. A given utility system could be faced with a peak electricity problem—if large numbers of solar energy systems relying on electrical back-up suddenly demanded energy from the central grid.

Active solar systems can also be adapted to use solar heat to drive a refrigeration cycle, so the systems can air-condition, as well as heat. Absorption chillers, using special refrigerant cycles, have been used for years; conventionally, they have utilized hot water or steam to run the process. Vapor compression and absorption cooling can take advantage of solar heat, rather than use steam from conventional sources. However, all the cooling technologies require specialized solar collectors, so that higher temperature (180–200° F.) heat can be supplied for the cycle.

The most advanced application of solar technology in the United States is the solar water heater. Like space heating, the solar water heaters use flat-plate collectors and pumps to circulate the solar heat into storage tanks. Instead of heating space, the units heat domestic water. Only a few years ago, it was thought that this technology would generate a multi-billion dollar industry by the mid-1980s—since water heating uses a lot of energy. However, regulatory requirements and manufacturing costs have driven the costs of the water heaters up to $3000 to $4000 per commercially installed household system. Even with generous federal incentives, such as a 40 percent IRS tax credit, the market has failed to develop, and two of the largest manufacturers of components have dropped out of the market. Exxon's *Daystar* collector division was recently sold to a Texas company, and Grumman's solar energy subsidiary was all but closed down in the opening months of 1981.

Given the ever-rising costs of active solar heating and cooling systems (up to $20,000 per home), it seems unlikely that the market will prove to be significant. And, as we have seen, the water heating heat pump is a substantial competitor to solar water heaters. These heat pump systems have the additional advantage of reducing peak electricity demands, since they require less than half the energy of electric resistance elements. In some new home installations, active space and water heating systems may be a promising means of saving energy, but the best alternative is still the passive approach—combined with sensible design.

A recent study by Christopher Flavin of the Worldwatch Institute in Washington, D.C., suggested a national goal of building (or remodeling) five million structures to be "climate sensitive" by 1990 and to "have such structures dominate the building market during the nineties." This, he says, would save an more than 10 percent of the energy needs of future buildings. This would reduce the fuel requirements of current buildings by one-third by applying reasonable retrofit approaches.

ELECTRICITY FROM THE SUN

Throughout history, people have used solar concentrators to generate not only heat, but even fire. The Greeks were not only active in the field of solar design for buildings and cities; there is also evidence that they attempted to concentrate sunlight with reflecting mirrors. According to the Greek physician Galen, in his *De Temperamentis,* Archimedes used mirrors to set fire to a Roman fleet besieging Syracuse in 212 B.C. This was not mentioned by the Greek biographer Plutarch, who, however, notes that the vestal virgins lit their sacred fires in seventh century B.C. with sunlight focused by goblets of reflective metal.

In the eighteenth century, the famed French chemist Lavoisier built a glass lens fifty-one inches in diameter and filled with alcohol to concentrate sunlight in order to melt metals. He succeeded in melting platinum—at 3190° F.— with his solar concentrator. In the nineteenth century, several European and American inventors built engines that used concentrated solar energy. Among them was John Ericsson, a talented inventor who is best known for his design of the *Monitor,* an ironclad ship used in the Civil War. Ericsson used a rectangular collector made of silvered window glass to concentrate the solar heat, and designed the device to "track" the sun across the sky. In 1884, he described his work in the British journal *Nature:* "Practical engineers, as well as scientists, have demonstrated that solar energy cannot be rendered available for producing accurate curvature on a large scale, besides the great amount of labor called for in preventing the polished surface from becoming tarnished, are objections which have supposed to render direct solar energy practically useless for producing mechanical power."

In 1878, Ericsson wrote a letter in which he reluctantly concluded that "the fact is . . . that although the heat is obtained for nothing, so extensive, costly, and complex is the concentration apparatus that solar steam is many times more costly than steam produced by burning coal." Ericsson's hopes for supplying his engines to benefit what he had termed the "sun burnt regions of our planet" were never realized, and he ingeniously converted his engines to run on coal and gas, selling more than fifty thousand of them on the international market to turn the sun-inspired failure into a commercial success.

Others followed in Ericsson's footsteps; in the early years of this century, solar projects using concentrating collectors were built in southern California and in Egypt—in both cases, to produce energy for irrigating arid lands. However, these were not commercially successful.

Today, there are several international programs that utilize the enormous energy of concentrated sunlight. Under the leadership of Felix Trombe, the French have designed and operated a ten-story parabolic solar mirror, which has been feeding a thousand kilowatts of sun-generated power into the power grid for several years. In the United States, several projects are underway using "power towers," as they are called, to make electricity. In all of these systems, a field of mirrors called heliostats reflect sunlight to a receiver mounted on a tower—the intense energy generated is used to heat a fluid, which can then produce steam for the operation of an electrical turbine generator. A central receiver system is operating at Sandia Labs in New Mexico for test purposes, and a $140 million system covering 130 acres is under construction in Barstow, California. This project, costing about $10,000 per kilowatt of installed electrical capacity (peak), has dominated the Department of Energy's budget for power towers for several years.

Power towers are a more efficient way of using solar energy to produce electricity than are "farms" of concentrating collectors because the energy is transported as light rather than heat. Parabolic reflectors focus sunlight in a line concentrating it up to thirty to fifty times. The receiver contains a fluid in a glass-lined tube that has a selectively absorbent surface and is located on the line of focus; it can reach temperatures of 572° F. Effective day-round performance requires that the collector track the sun on at least one axis. A variant of the parabolic trough collector uses tracking Fresnel lenses, which are less sensitive to tracking errors than are mirrored systems.

These applications include industrial process hot water, steam for industrial applications or electrical production, and space cooling. Annual efficiencies in existing demonstration projects are only eight to thirty percent. Nine percent of the U.S. end-use energy demand is for low temperature industrial process heat (less than 550° F.) This is an ideal market for trough or other collectors. Parabolic trough collectors will probably be able to satisfy a good portion of this demand as production costs decrease and techniques for integrating the systems into industrial processes improve.

Centralized systems can produce large amounts of energy in areas that receive high solar radiation and where

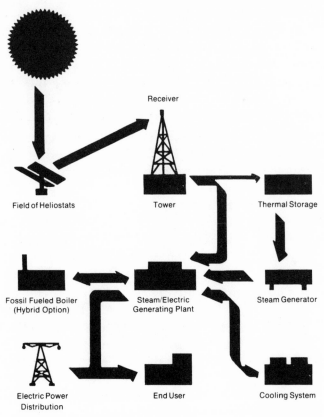

Figure 13. Solar Central Receiver System

other land use options are limited. However, decentralized collection systems have greater flexibility because they can operate in a modular fashion—adding or deleting units according to demand. Energy use at the site of production reduces transmission losses that occur when electricity is distributed over long distances. It also makes possible the salvaging of waste heat, which cannot be economically transported over long distances. Decentralized systems also reduce the vulnerability to large-scale energy shortages.

Proponents of central receiver facilities predict that we will see some three to four hundred plants producing about a hundred megawatts of electricity apiece, during solar peak periods. During the night, the electricity stored in batteries will be available for use. However, we question the centralized approach, since the additional costs of batteries, transmission lines, and other factors may render this central technology an over-priced, low-efficiency solution to our worsening energy situation. On a local basis, however, the use of concentrated solar energy for making electricity may be a significant energy technology—in sunny locales. We might also learn from John Ericsson's experience in the last century, and find that parabolic dishes might be better used to see if there is intelligent life in the universe.

OCEAN THERMAL AND SOLAR POND POWER

The tropical and subtropical oceans are massive natural storage basins for solar energy. Tropical currents and the sun warm the ocean surface, while, deeper down, the polar currents continuously chill the bottom water. The means to utilize the differential to make electricity have been designed and the concept has all the advantages of a vast heat engine. The projected system would consist of a power plant at sea, with a mammoth vertical pipe for sucking cold water from the ocean bottom. In the plant, warm surface water would be pumped into an evaporator containing a substance like freon, which would turn to vapor and expand through a turbine. The cold water from the ocean bottom would then condense the freon back to its original liquid state—and so on. The electricity thus produced would be transmitted to the continental power grid by an underwater cable. The system is called Ocean Thermal Energy Conversion (OTEC), and early estimates for its future suggested that such plants located in the Gulf of Mexico could produce several thousand gigawatts (a gigawatt is 1000 megawatts)—more than enough to provide U.S. energy demand for an entire year. Those estimates were optimistic. Further research and engineering studies have shown that there are only a few sites where, for technical reasons, OTEC plants could be used. Yet enthusiasm for the project has not abated. In February 1981, Governor Harry Hughes of Maryland announced his intention of turning his state into a construction center for OTEC equipment, envisioning the creation of as many as twenty-five thousand new jobs. Explaining the technology to reporters, he said, "It's almost perpetual motion."

In addition to the technical and engineering difficulties to be overcome, however, an OTEC plant poses a more serious drawback; it would be an enormous installation—a sitting marine duck. The basic principle of OTEC, however, could be used on land in smaller, less vulnerable sites: for example, salt ponds, either natural or manmade. Typically, in a salt pond the water in the top layer is low in salinity and density, while the bottom layer is very saline and dense. Solar radiation penetrates the water and is converted to heat; the heat is then trapped in the bottom layer's dense, salty water and "blanketed" by the less dense, less salty layer of water above. The pond becomes both the collector and the storage medium. Such a pond can be used straightforwardly to produce heat—such as one built by Miamisburg, Ohio, to heat an outdoor swimming pool and an adjacent recreation building. Even in February, the bottom temperature of the pool is eighty-three degrees.

The salt pond can also be used to run the saline layer through a heat exchanger and on to a system that produces electricity. The Israelis were first to experiment with this concept, building a 150 kilowatt power station in 1979 at Ein Bokek on the Dead Sea. Now the State of California, the U.S. government, and an Israeli firm are studying several American sites including Great Salt Lake and the Salton Sea. In a typical solar pond power plant, hot brine is piped to the evaporator of a heat engine, where the brine's energy converts a fluid of very low boiling point from liquid to vapor. The vapor then expands through nozzles of a turbine, turning the turbine wheel and the generator that produces electricity. Finally, the vapor is led to a condenser, where cooler water from the top of the pond cools it back to a liquid state, and the cycle begins all over again. The idea works: in Israel, lights are burning and motors turning with electricity generated in just this way. The special turbines were developed by Ormat Turbines, an Israeli company started in 1964.

During the developmental work on the Israeli pilot plant, Ormat's president, Lucien Bronicki, journeyed to California for a special meeting with Governor Edmund G. (Jerry) Brown, Jr. They discussed California-Israeli cooperation in alternative energy. Bronicki proposed that the 360-square-mile inland Salton Sea be tapped for its brine and appropriate desert location for a solar pond demonstration. In Bronicki's subsequent meetings with the Southern California Edison Company and other groups he discussed plans for an

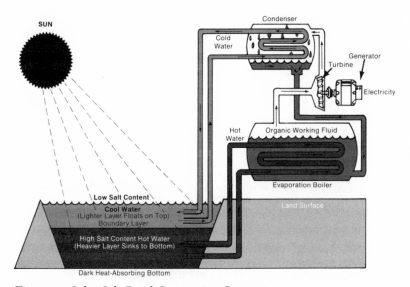

Figure 14. Solar Salt Pond Generating Concept

ambitious development project to be completed in the 1980s.

Feasibility studies for the Salton Sea project are to be completed by 1981; design and construction on a prototype plant, using a diked pond covering about four-tenths of a square mile, are to begin in 1982. If all goes well, construction of the modules comprising the full-scale six-hundred-megawatt plant—using a total of forty-six square miles—will begin in 1985. How much would it cost? No one yet has figures for the modules, but Ormat estimates the 5-megawatt prototype would cost about $10 million and deliver electricity for about seven cents a kilowatt hour—the same as electricity costs from a coal-fired plant. Playing the prediction game, Ormat's scientists estimate that the western United States has the potential for producing more than twenty thousand megawatts through solar salt ponds, and Bronicki has visions of solar salt ponds supplying one-third of Israel's energy by the end of the century.

PHOTOVOLTAIC ELECTRICITY

Another solar technology that shows great promise makes use of photovoltaics—or, solar cells. They operate on the principle that photons from the sun, when they strike a cell made up of both a positively and a negatively charged slice of silicon, create electricity. Since the beginning of the space age, solar cells have been used—in spite of their enormous cost—to power orbiting satellites. The costs in those earlier days were around $500 per peak watt. After a great deal of research, which began in earnest with the oil embargo of 1972, that figure has come down to $10 and is likely to drop even further, making photovoltaic arrays for producing electricity competitive with oil-fired plants by the mid-1980s. The kinds of improvement being made in this technology are illustrated by the "violet cell" developed by Dr. Joseph Lindmayer in 1972. This breakthrough made it possible for the cell to utilize more light and increase efficiency. Current research is directed at lowering the cost of silicon, finding less expensive manufacturing methods of silicon crystals, finding new materials, and trying new designs that will reduce the amount of silicon needed. For example, in 1978 the California firm Varian Associates combined gallium arsenide with silicon cells and achieved an efficiency of 28 percent in converting light to electricity—the highest ever reported for solar cells.

There are, of course, many obstacles to be overcome before photovoltaics will be practical for commercial use. Producers are reluctant to scale up their manufacturing efforts until sufficient demand has been demonstrated. But demand depends on the low prices that mass production would bring. One side or the other must be stimulated, and most people in the field recommend massive purchases by the government: a photovoltaic "Manhattan Project." In any case, current levels of funding are insufficient for this kind of massive stimulation. (The Energy Department had been working on an ambitious development program, until President Reagan's budget-cutters axed most of it in 1981. The goals had called for developing a photovoltaic industry capable of delivering 50,000 megawatts of solar cells—at peak energy production—per year by 1990, at costs between 14 and 42 cents per peak watt.)

In spite of setbacks such as budget-cuts, progress with photovoltaics is being made. New photovoltaic designs provide a photovoltaic roof shingle for homes (from the Atlantic Richfield Company), a multilayered sandwich of silicon wafers that are illuminated from the edge (expected to reach an efficiency of up to 30 percent), and cells of pure silicon that are expected to reach 20 percent efficiency after some improvements are made.

While researchers work on making photovoltaics more economical, the projects using baseline technology show that photovoltaics work today. A $27.5 million federal scheme has built systems under the Federal Photovoltaic Utilization Program (FPUP). The first fifty-three installments cost $500,000 and were built for the Forest Service, the Navy's Material Command, the Indian Health Service, and the Tennessee Valley Authority, all at remote locations. One of the locations is Schuchuli, Arizona, an Indian pueblo of ninety-six people. The PV supply will provide them with 3.5 kw for lights in all houses and the community house, a washing machine, a sewing machine, a water pump, and fifteen small refrigerators.

Another photovoltaic installation that has received widespread notice is a 283 kw unit at the Phoenix Sky Harbor International Airport. At peak generation it will provide enough electricity to power forty average-size homes in Phoenix. It is being built by a team composed of Arizona Public Service Company, Motorola's Government Electronics Division, the City of Phoenix, and the Arizona Solar Energy Commission. It will use 7200 concentrator modules on thirty large arrays. The total array will require ten acres at the airport and will power half of the south concourse of a new airport terminal.

Presently, however, there are several major obstacles to the rapid commercialization of photovoltaics. These include a projected decrease in the production of high-purity silicon crystal, the lack of ready capital for expansion of the industry, the threat of foreign competition for markets (Japan, France, Germany, and Italy all have active PV research and development programs), and unresolved questions about the socioeconomic and environmental impacts of the rapid development of new technology.

A shortage of silicon is the most immediate problem. New processes now being developed may not be ready in time to relieve the shortage. No solutions have been offered by government researchers to date, and silicon producers are reluctant to invest the massive amounts of capital required to assure production until they have a more solid indication of demand. One way of solving the supply problem will be to use a cheaper metallurgical grade of silicon. This approach has been incorporated into several research programs, including Crystal Systems, which has achieved a cost production rate of $3 per kilogram. ARCO Solar has also proposed that silicon manufacturers produce a "solar grade" of silicon, which would be more pure than metallurgical grade silicon but less pure than semiconductor silicon.

Other unresolved questions that impede the advance of photovoltaic systems include the environmental problems of toxic substances created in cell manufacture and use—along with the social impacts of the rapid development of the new technology. The Solar Energy Research Institute is beginning to study environmental hazards, but little information has been available until recently because both materials and processes have been in such a state of flux. The manufacture of photovoltaic cells is a highly energy intensive process. Under existing technology approximately 7000 kilowatt-hours are required to manufacture cells with a peak output of one kilowatt. This means the device must operate for about four years to "pay back" the energy consumed in making it. Although cheaper, less energy intensive methods for producing these devices are being developed, a massive expansion of the industry is likely to require a major use of energy from other sources. In the long term, of course, the goals should be to power the solar industry with renewable energy sources exclusively.

Dr. Joseph Lindmayer, developer of the violet cell and president of Solarex, Inc., expresses the following goals: "Beginning in 1973 the groundwork for a basic industry was laid, and by 1980 solar panels could already be produced at $10 a watt. Production capacity was a few megawatts (i.e., thousands of kilowatts) per year. Direct competition with the electric utility will fully develop when the cost will be about a dollar a watt, and this figure could be achieved in the 1980s.

"It is of great importance to society that alternative energy sources have net energy gains. Another idea, which I have proposed, is the creation of a solar breeder. This facility would use electricity from banks of solar cells to produce more solar cells. No outside source of power would be needed. And a solar breeder would show a net energy gain in about a year.

"We have a great need for such power gains, since civilization can only flourish when an energy surplus feeds it. In this respect, photovoltaics are truly ideal."

One of the most intriguing, and controversial suggestions for photovoltaic applications is the solar power satellite (SPS), proposed by Dr. Peter Glaser. In Glaser's scheme, massive discs (over a half-mile in diameter) covered with photovoltaic cells would be positioned in stationary orbits 22,500 miles from earth. Power collected on a twenty-four–hour basis would be transmitted to earth using microwaves or laser beams. Although such a massive system could probably be built, it would represent a vulnerable target in much the same way that the ocean thermal plants would. Nonetheless, Dr. Glaser makes a persuasive case for the SPS: "Solar cells, with only minor modifications, could be used on Earth or in space. And curiously, SPS development could encourage distributed solar technology because photovoltaic research for the SPS could also be beneficial for terrestrial photovoltaic applications. Most likely, I suspect, strong central planning will be required for both dispersed and centralized solar technology applications to succeed. Awareness of these technologies and their potential by individuals, communities, regions, and countries will surely differ at various stages of technology development.

"The SPS could provide not only the impetus for peaceful cooperation among nations because all can share the limitless resources of space, but it could help us achieve the inevitable transition to renewable sources of energy, inevitable, that is, if our advancing world civilization is to endure and mature."

Regardless of the merits of Glaser's system, it seems unlikely that the present attitude of the federal government will permit any such large-scale application of photovoltaic energy—whether on earth or in space. The development of this technology will probably proceed at a more leisurely pace, without the impetus of substantial federal funds. Given the interest and investment of many and diverse economic entities—including the large oil companies such as ARCO and SOHIO—photovoltaic energy development will surely proceed somehow, and many applications will be ready for commercial use in the decades ahead. Dr. Reinhard Stamminger of the Monegon Corporation advocates that oil companies and public utilities take a more active interest in photovoltaic research and development—spending a fraction of their income today to build a strong, renewable energy base for tomorrow. He believes that utilities should pursue the decentralized use of photovoltaic power and has stated that "One strategy for electric power utilities to follow could be the early implementation of distributed photovoltaic systems. To the extent permitted by Federal and State regulations, the costs for these systems would be included in the rate base, as long as this is done before the concept of competition in electric power generation becomes widespread. Electric power utilities would then provide the required stimulus to the rapid introduction of photovoltaic systems. The higher costs of the photovoltaic systems in the early years would be offset by lower costs in later years; and in any case, cost averaging may be acceptable. Total utility revenues for 1980 are estimated at $90 billion. Allocating one-tenth of 1 percent of this amount to the implementation of photovoltaic systems would provide the necessary stimulation for the photovoltaic industry to reduce costs to a competitive level during this decade and would establish the utilities as a major owner and operator of photovoltaic systems."

Photovoltaic systems are also ideally suited to residential and other on-site applications. On-site locations, connected to the utility grid, could take advantage of buy-back arrangements and use utilities' production power at night. Private citizens could combine arrays for clusters of homes; utilities could use arrays on a neighborhood or district scale as load levelers, or for remote mechanical tasks such as running irrigation systems.

It seems certain that photovoltaics can and should play an essential role in a dispersed and diversified energy system, converting as quickly as possible to the use of renewable energy sources. And it seems certain that the federal government should seriously rethink its role in choosing between solar technologies so that they encourage those that have the highest promise—not only as energy systems but also as strategic elements.

POWER FROM MOVING WATER

The sun, that great engine of the solar system, also drives the hydrological cycle of the Earth and, in conjunction with other forces, creates the waves of the ocean, which possess tremendous energy, as anyone who has been caught by one and "boiled" up the beach knows, and have occupied the minds of inventors for many years. There are all kinds of devices for harnessing this marvelous energy but generally they do not work very well. It is one thing to power a buoy: attach a tube to a float, and in each wave trough, water will flow up into the tube; a valve will keep the water in the tube when the waves crest until there is enough pressure to generate power. It is entirely another thing for waves to produce an amount of electricity substantial enough to serve a community.

In the United States, the most likely place for the use of wave-energy is the Pacific Northwest, where the Oregon and Washington coastlines enjoy the most consistent wave conditions. Potential wave power along this coast is estimated at 87,000 megawatts, as opposed to about 10,000 along the entire American coastline of the Gulf of Mexico.

The drawbacks to harnessing the power of waves are

technical, economic, and environmental. The devices must be very large in order to be successfully moored. Oil rigs may serve as models, but it seems likely that the costs of mooring wave-energy plants will prevent the business from becoming economic. Transmission cables are an additional problem. And funding of research by the federal government is slight —perhaps properly so: about $1 million a year, and there are no large scale development plans for the future. Environmental problems enter the picture because altering wave action would alter the shoreline (as the Army Corps of Engineers well knows), and the shorelines would be further industrialized.

HYDROELECTRIC POWER

The power of falling water is an important source of industrial energy and has been since the beginning of the industrial revolution. Between the thirteenth and the nineteenth centuries, almost every village and town in Europe with access to running water was utilizing hydro power for everything from tanning leather to grinding grain. The early water mills and windmills averaged about five to ten horsepower per installation, and they ushered in the industrial revolution.

Today, hydroelectric plants supply a little under 4 percent of the overall energy use in the United States—slightly more than the contribution of nuclear power. Although hydro power plants are generally smaller than coal and nuclear plants, some are quite large, such as the 700 megawatt plants on the Columbia River in the Pacific Northwest and the 2,000 megawatt plant on the Niagara.

In the mad rush to develop ever bigger centralized power plants, many of America's smaller hydro facilities were abandoned. This trend accelerated in the last few decades, when electricity from fossil fuels was far cheaper than electricity that the small plants could provide. Recently, however, a sharp reversal has occurred, and an entirely new trend in the direction of small hydro plants has developed. Small hydroelectric power systems are water-electric power systems, up to 30,000 kilowatts (30 megawatts) in size. The hydraulic "head" is comparable in most cases to that found in larger hydro installations, but a smaller water flow restricts electrical capacity. Conventional, but smaller, turbines, generators, governors, and control equipment are used in small hydroelectric plants. Small hydro power facilities are used in many parts of the world, and there are extensive installations in Europe. The People's Republic of China is the world's leader in small and micro hydro power plants, with over 90,000 installations providing more than 5,400 megawatts. The Chinese small hydro plants are quite decentralized in nature and are either not grid-connected or feed power to local grids for small industries associated with rural communities.

Currently existing hydroelectric power facilities generate 63,702 megawatts. Of this total, 2,957 megawatts are produced at small-scale sites (05–15 megawatts); 1,517 megawatts are produced at intermediate sites (15–25 megawatts); and 59,230 megawatts are produced at facilities larger than 25 megawatts. There are over 5,600 small-scale dams in the United States either generating power or with the potential for a generating capacity. The values for small-scale capacity and generation represent about 5 percent of the nation's current installed hydroelectric capacity and energy, according to the Army Corps of Engineers. In addition, a recent Army study indicates that over 16,000 megawatts of electricity can be generated from new small hydro plants.

The distribution of existing small power production facilities is extremely variable, and nearly all regions of the country have the potential for incremental energy development. Currently, most small scale hydropower installations are in the Northeast and the Lake Central regions of the country. The underdeveloped hydroelectric potential at small-scale sites is widely distributed but appears to be greatest in the

Pacific Northwest, Lake Central, and Northeast regions.

The Army Corps estimates of future potential are only approximate and do not take into account classes of hydro projects such as those associated with canal drops, pipelines, pressure breaks, and other facilities that are part of municipal and district water supply systems. These sites are becoming increasingly attractive as the economics of energy production change dramatically, and many such projects are under study. The federal government has recognized the importance of such projects and has written regulations granting exemption from the Federal Energy Regulatory Commission licensing procedures for manmade conduits that generate hydroelectric power. For projects up to 15 megawatts, ceusing exemptions have been given under most circumstances.

There is nothing particularly complicated about small hydro technology, which employs various kinds of turbines to convert the mechanical energy of moving water into electricity (or motive power). Companies like the James Leffel Company of Ohio began making the necessary components in the 1800s. As is the case with many of the other technologies discussed in this book, small hydro technology is rapidly being developed by other countries. Leffel, for example, was recently purchased by a Scandinavian manufacturing firm called Nohab Tampella. A recent United States advertisement for the firm noted that "An important part of Leffel's activities is in the field of rehabilitation. Leffel's total service program includes expert inspection of existing installations, complete rehabilitation of abandoned sites, experienced repair and installation of mechanical-electrical parts plus complete supervision of field service personnel."

The opportunities for increasing U.S. energy efficiency through expansion of small hydro facilities may yield a net result of supplying about 10 percent of the nation's electricity. As with other small technologies, the long lead times and regulatory delays of coal, synfuels, and nuclear plants will not substantially impede the growth of small hydro plants. By nature, this is a community-based and sensible technology for development, with the added bonus of a "free" renewable energy base.

The only major use of the sun's enormous energy—outside of agriculture and forestry—by any society has been the harnessing of the rivers for motive power and electricity. As the newer small hydroelectric plants begin to be used more, hydroelectric power will play an increasing role in our energy economy. Yet the other prospective direct solar technologies describes in this chapter offer the potential of limitless energy for mankind, if we can learn to use them. The major contributions from solar energy will come from the intelligent application of sound architectural principles and passive solar concepts. Nonetheless, we will need to experiment with and develop the higher technology solar applications—ranging from solar ponds to photovoltaics.

10. Agriculture, Wind, and the Heat of the Earth

No SCIENTIST—or any thoughtful person—fails to recognize the primacy of the Sun in the material part of our universe. And it should be evident that no energy system that we have the slightest dream of capturing and mastering comes from any place but the Sun. The asphalt road that you drive over is made up, largely, of well processed dead plants that lived because of the sun. They died hundreds of millions of years ago and, through a process not yet entirely understood, turned into petroleum. And billions of years before that—also through a process that is not yet thoroughly understood—the Earth and the other planets formed out of a swirling cloud of gas, a cloud consisting of that which was not coalescing into the Sun itself. When will we truly understand the fantastically complex series of causes and effects by which the energy and the cycles of the Sun affect our changing climate on this planet—or the far more variable matter called weather? In this chapter, we look at three promising energy sources derived from the sun, but not usually considered "primary" solar energy sources: biomass energy, from plants and crops; geothermal energy, from the stored heat of the earth; and wind energy, a solar derivative. With proper management, and the development of new technologies, these sources could provide an endless supply of energy.

BIOMASS: PHOTOSYNTHESIS AND ENERGY

All biomass energy results from the single most important chemical process on Earth: the photosynthetic effect of sunlight on green plants. Photosynthesis breaks down water and carbon dioxide gathered by the plant into sugar—or, carbohydrate, which is a composite of carbon, hydrogen, and oxygen—and pure oxygen. The miracle of photosynthesis, discovered two hundred years ago by Joseph Priestley, was beautifully described by the eminent botanist Donald Culross Peattie in *Flowering Earth:* "As each bullet [of light] hits an electron of chlorophyll it sets it to vibrating, at its own rate, just as one tuning fork, when struck, will cause another to hum in the same pitch. A bullet strikes—and one electron is knocked galley west into a dervish dance like the madness of the atoms in the sun. The energy splits open chlorophyll molecules, recombines their atoms, and lies there, dormant, in foods." Plants store the food energy supplied by photosynthesis in such quantity that they in turn supply animals with food energy as well. Until the widespread use of fossil fuels, the photosynthetic energy transfer sustained both agriculture and industry. Today's fossil-based economy relies on the daily photosynthetic pulse for food and fiber.

In the United States, the combustion of wood provided almost 90 percent of all energy requirements until the end of the Civil War, in 1865. In the last century, the clearing of forests for agriculture and for firewood resulted in widespread despoliation of native ecosystems—later results of which include massive soil erosion and conditions such as the dust bowl of the 1930s. Even today, the combustion of wood is a significant source of energy, as the growing sales of wood stoves attest. Biomass sources of energy will continue to be major contributors to our energy economy, but we must use great care in the selection of technologies and the allocation of land to this task.

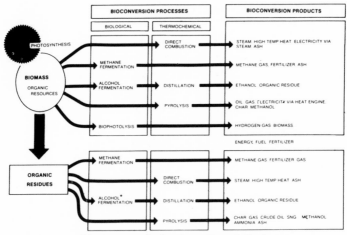

Figure 15. Biomass Conversion Processes

The efficiency of converting biomass to energy is abysmal —only 1 or 2 percent, compared to the efficiency of photovoltaics, which can go as high as 35 percent. Moreover, it requires large amounts of land to capture solar energy through photosynthesis. The United States is a world leader in superbly productive agricultural land. Furthermore, current U.S. agricultural and forestry practices produce enormous quantities of wastes and residues. Various estimates of the biomass potential suggest that we could, by the year 2000, derive between 7 and 20 quadrillion BTUs—the equivalent of as much as 10 million barrels of oil—per day from this source alone.

Currently, the United States produces the equivalent of about a million barrels of oil a day from biomass—chiefly from using forestry wastes to produce process steam and cogenerated electricity. It is not very economical as an energy-producing system because the costs of transportation of forestry wastes are generally attributed not to the system itself but to the cost of the lumber. But in Eugene, Oregon, there is a city-owned public utility that has expanded its existing wood-fired electrical generating facility, which is currently seeking forest slash to make electricity at the lowest cost to its customers. It all depends on whose economics you are subject to. And economics can be changed rather simply—by rewriting tax codes and stimulating markets.

Instead of growing soybeans for an uncertain market controlled by Washington, a farmer might grow grain purely for the purposes of producing alcohol. Presently the economics of biomass energy are tenuous, given the *relatively* low cost of conventional fuels. But this will change. A current example of cost-effective farming would be growing grain to produce alcohol, which would leave a residue of high-protein cattle feed. The cattle produce food for human consumption; their waste produces energy, if converted to methane, which in turn produces process heat to run the alcohol distillery; then the alcohol produced in the first place drives the tractors and machinery of the farm. It takes organization, and organization is the opposite of entropy. No farm, house, or any other system can ever be altogether energy–self-sufficient and as productive (or consumptive) as it originally was without outside inputs. Entropy is inevitable, but waste can be lessened through the use of efficient technology, design, and planning: for example, the farm described above, which aims for a 90 percent return on the energy-in/energy-out equation, getting every last scrap of usable BTUs out of every bit of energy produced. This kind of organization requires thinking that is both precise and small: thinking incrementally. It is not what economists are accustomed to doing. Like architects, they like grand solutions, of which—we are persuaded by their efforts to date—there are none.

A far more practical alternative than the massive gasohol factories proposed for America's cornbelt is a small community-scaled alcohol still, capable of meeting regional energy needs and operating on local crop residues and a portion of local agricultural production. One such plant was put into operation in October, 1980, by the Janss Energy Corporation at Jerome, Idaho. The still is classified in the "tiny" range by the U.S. Energy Department, since at full production, it makes only about a half million gallons of alcohol a year.

According to Ed Janss, founder of the alternative energy company based in Los Angeles, the Jerome plant was put into operation only eighteen months after he conceived of the idea—a remarkable accomplishment, considering that most energy projects take a minimum of five or six years to construct and license. He predicts that the $1.2 million plant is only the forerunner of a whole new generation of efficient alcohol technology on the local level. There really is no need to make anhydrous-grade alcohol (200°) for gasohol production. Producing lower-proof alcohol (160–170°) that can be used directly in cars, trucks and tractors is a better solution. This saves a substantial part of the energy in the production process required to make anhydrous-grade alcohol.

We foresee even more efficient developments for rural areas, such as this small Idaho valley. Local farmers would use "batch-process" stills to partially upgrade alcohol from their wastes and crops, then larger stills would send tank trucks into the countryside to collect the waste and take it to more efficient central facilities for conversion to fuel-grade alcohol for direct use. The direct use of alcohol for fueling vehicle fleets may be one of the most efficient ways to make use of the fuel. The state of California is currently testing small fleets of specially designed Volkswagen Rabbits and Ford Pintos to run directly on alcohol.

A prodigious number of useful items can be manufactured from biomass, just about everything that can be made from coal. And it is precisely coal that biomass must compete against as a source for heat and for gaseous fuels. In some instances, it is economically and technically competitive, in others it is not. But as Battelle Pacific Northwest Labs points out, "biomass is a renewable source and coal is not."

However, the conversion of coal into methanol may prove to be a major transition to the development of an "alcohol economy" for the United States. This synthetic fuel technology offers the potential for developing a major alcohol industry. Already, companies such as W. R. Grace and Company are planning major methanol projects; the Grace project, located in northwestern Colorado, would produce five thousand tons of methanol per day from regional coal deposits.

According to *Fuel Alcohol,* the report of the U.S. National Alcohol Fuels Commission, issued in 1981, a variety of coal grades can be used for conversion to methanol, including lignite—a plentiful, high-moisture coal found in North Dakota, Montana, and Texas. The methanol synthesis process virtually eliminates gaseous sulfur emissions, because the process requires desulfurization of the coal gases. According to the commission, "This means that methanol plants can use a higher-sulfur, lower-quality coal and avoid price competition with steam coal users who prefer higher-priced, low-sulfur coals." This would be an important factor in eastern states that have large reserves of high-sulfur coal.

Over a billion gallons of methanol are produced annually in the United States, but this is almost entirely made from natural gas. In the future, as more uses are found for alcohol as a substitute fuel, coal-derived methanol will prove to be a major domestic energy resource. It will have the advantage of flexible pipeline transportation to the point of use—be it a utility boiler, an industrial plant, or a vehicle fuel tank.

Biomass energy conversion systems do have several obvious problems associated with them. One glaring fault would that if farmers grow plants to produce energy they are not growing food for people to eat. A balance must be worked out. On the other hand, there are a great number of agricultural wastes involved in putting food on the table, from corn-stalks to wastes from such steps in the agricultural chain as food processing and canning. There are also agricultural products and wastes that have too much cellulose in them to be useful for food. It is in such matters that an economic calculus of biomass energy production lies.

One of the most practiced ways to use biomass is burning or oxidizing it to produce heat. California's Pacific Gas and Electric Company has just signed an agreement to buy electricity from what will be the world's largest biomass plant. The plant will burn dried pellets of agricultural waste, such as grape clippings, sawdust, and cotton waste, to produce some 50 megawatts of power in Madera, a town in the San Joaquin Valley. The plant, privately built and planned for operation by 1982, will also install a generator to produce steam that will dry the agricultural waste pellets—a variety of cogeneration that PG&E hopes will help satisfy additional power needs. There were times not long ago when one could overhear utility and energy people ridiculing the use of such things as grape clippings and sawdust to make electricity. No more. Besides making heat from this material—what you might call preformed coal, or, non-fossilized fuel—we can use bacteria to turn it into gases or liquids.

Low BTU gas is a revived technology that has in the past achieved levels of sophistication capable of sustaining petroleum-starved countries during wartime. Low BTU gas is competitive with even deregulated natural gas, but it suffers from a lack of proof of technological long-term reliability. Still, its small-scale versatility and probability suggest a successful future—especially in applications like off-road vehicles and farm equipment. Low BTU gas is particularly good for providing process heat and steam or for electrical power generation. But it has to be consumed near the source of production. It is a viable, decentralized fuel.

Medium BTU gas can be transported moderate distances; it requires rather larger centralized operations and is usable in synthesizing higher energy forms, such as synthetic natural gas or chemical feedstocks. The production of medium BTU gas will not be competitive with coal in the near term, and, since it takes such a large operation to be economically efficient, it is not likely that such plants will be effective in the use of cellulosic biomass, such as forestry wastes.

Producing methane, by fermentation of cattle wastes or sewage, appears to be competitive with the production of deregulated natural gas—but, in the case of cattle wastes, only on a spread of 100,000 head or larger. Yet 75 percent of confined cattle are on farms, ranches, or feedlots of less than 1,000 head. Dairy farmers seem to be taking the lead here in devising unconventional systems, and this technology, abandoned by the Germans and the French after World War II, will soon provide reasonable pay checks.

Ethanol seems to be the most advanced technology at present, and plans have been announced by a variety of private companies to build about 340 ethanol plants in the 1980s, ranging in size from small operations like the Janss plant in Idaho to huge facilities, capable of producing 100 million gallons of fuel a year. The process now relies on grain to a large extent, which raises the issue of food versus energy competition—although there are a number of other potential feedstocks. Table 4 shows the potential of a number of industries to convert to ethanol production over the next few years.

In the future, a number of processes other than fermentation may prove economic, including the conversion of cellulose directly to ethanol. Use of special acids and enzymes to break down the cellulose molecule is a target of many industrial research efforts. Cellulose is an abundant plant material that would free more vital agricultural crops for food consumption.

LOCAL USES

An important, overriding goal for the nation should be the logical introduction of alcohol fuels into the fossil economy. Small facilities should be encouraged by government poli-

Table 4. Capacity of Industries to Convert to
Fuel Ethanol Production

Industry	Potential Annual Capacity through 1985 (millions of gallons)	Estimated Capital Cost through 1985 (millions of dollars)
Distilleries	95	22.6
Breweries	215	257.0
Sugar Beets	100	98.2
Corn Wet Milling	93	68.9
Sugar Cane	8	6.5
Cheese Whey	44	242.2
Potato Processing Waste	11	26.2
Citrus Processing Waste	15	33.0
Total	581	754.6

cies, including a careful analysis of federal programs that subsidize behemoth facilities and discourage community-based plants. According to Bill Holmberg of the U.S. Department of Energy, a primary goal should be to make local communities more self-sufficient through renewable energy technologies: "Water, topsoil, and local independence are the criteria we should utilize to gauge the merit of these technologies."

In hearings of the Senate Committee on Governmental Affairs, Admiral Thomas Moorer, formerly chairman of the joint chiefs of staff, cited the need for alcohol production by the farm communities of the United States, which, he said, "can make an important contribution to national resiliency." Moorer noted that, "the nation should explore policy options in this sector, such as including supplies of high-grade liquid fuels for farm communities in all fuel allocation contingency plans; encouraging farmers to maintain petroleum fuel reserves on their farms by keeping storage tanks topped off;

reducing the dependence of the farm community on imported high-grade liquid fuels by producing fuel-grade alcohol within the community; and developing Federal or State
programs to encourage farmers and farm communities to
establish a national Defense Alcohol Fuel Reserve to meet
their needs during an emergency, as the Strategic Petroleum
Reserve will serve military, industrial, and urban communities. Such a program would not only provide the needed
reserve, but would also stimulate the advancement of the
alcohol fuels industry. This in turn will reduce our crippling
dependence on imported oil and will help sustain rural communities in the event of a prolonged or particularly damaging national emergency."

Admiral Moorer's suggestions reflect the possibilities for
integrating an alcohol fuel economy with the agricultural
sector. With local production of fuels, and the use of such
fuels in regional application, less initial energy will be required to produce net energy from alcohol. A major debate
has centered on the issue of energy "subsidy" to produce
alcohol. In other words, when all the fossil fuel energy inputs
to agriculture (fertilizers, farm fuel, pesticides, and so forth)
are added up, it appears that certain large-scale alcohol
plants—relying mainly on grains—cannot achieve a net energy balance. They cannot deliver more BTUs in alcohol
than the subsidized BTUs of petroleum in the process. But
this problem can be alleviated by attention to regional, decentralized applications of alcohol plants—along with other
biomass approaches as well. One such example of a high
efficiency regional process would be the production of
ethanol from biomass (or methanol from coal) and its use in
small cogeneration power systems. By using local fuels to
power modern facilities that could re-cycle waste heat, very
high efficiencies are possible at relatively low costs. In other
applications, such as fleet vehicles (about 10 percent of the
national population of vehicles) and aircraft, alcohol seems to
be a leading contender for high fuel economy. Bill Paynter,

the president of California's Union Flight Services, has tested methanol in aircraft and found that "above 10,000 feet, alcohol provides greater range per gallon of fuel than gasoline. With a high compression ratio and a turbocharged engine operating at 16,000 feet or higher, it is entirely possible that a formulated alcohol fuel would have twice the range of gasoline."

ENERGY AND WASTES

The operation of sewage treatment plants consumes enormous amounts of energy in the United States. Federal regulations promulgated by the Environmental Protection Agency require that treatment plants use recently developed technologies to produce environmentally acceptable effluent. However, this often means that older, more prudent techniques—like using methane from sewage to operate the plant's electrical generators—cannot be used, because the toxic chemicals of the newer systems kill the beneficial bacteria that could convert wastes to methane fuel.

A more acceptable technology is currently in use in a small, growing town called Hercules in northern California. In order to allow for continuing growth of population from its present 7000 inhabitants to as many as 23,000, the town realized it would have to increase its water treatment capacity. Their options were to increase the conventional plant in a nearby town, where its waste water was then being treated, to join with several other towns in a large regional system, or to build its own plant. The town chose a newly developed system called AquaCell developed by the small firm Solar Aquasystems, Inc. Feasibility studies were made in 1977 of the AquaCell system and of enlarging the facilities in the nearby town. The AquaCell system, it appeared, would be approximately $300,000 cheaper in annual operating costs and would use a third of the electricity that the existing plant used. The town leaders unanimously elected to use Aqua-

Cell. An ingenious structure, it consists of an inflated polyethylene greenhouse cover built over three treatment lagoons to stabilize water and air temperatures and to prevent excessive evaporation. Buoyant plastic-mesh ribbons are anchored in the lagoons increasing a hundredfold the surface area on which waste-digesting organisms live and graze. Thus, the organisms multiply faster and digest waste faster. In addition, duckweed and water hyacinths grow on the pond's surface, existing on nutrients in the waste water, screening out the sun, and thus inhibiting algae growth.

In the course of two days, the sewage flows through the three ponds receiving primary, secondary, and tertiary treatment; slowing the treatment to four or five days provides advanced treatment quality. Some nitrates and phosphorous remain. Some pathogens are killed in the long retention period; those remaining get trapped in the sand filtration system and can be killed with ozone as the water flows out of the main lagoon and into a clear well to be pumped away. The final volume of solids is estimated to be half that of a conventional plant's sludge. The plants harvested in the system can be composted by themselves or, along with the sludge, they can produce fertilizer. The plants, harvested alone, are high in protein and can be used for animal feed. In the process, some methane is produced, which could be used to fill the electrical requirements of the plant.

Another considerable advantage of the AquaCell system for the people of Hercules was that it did not have to be built all at once. Built on a modular principle, it can be extended gradually as the population climbs from the present 7000 to the projected 23,000. Taking their cue from the town leaders' enthusiasm for the AquaCell, many citizens installed low-flush toilets and restricted flow shower heads in their own homes—both of which are expected to reduce local water use by as much as 40 percent, resulting in a more concentrated waste-water flow that can be treated effectively by the AquaCell; a conventional plant, however, would not be able to process a highly concentrated waste-water flow.

Climate and land-use are among the factors inhibiting the widespread use of AquaCell system. They might not work efficiently in less temperate climates, where longer treatment periods would be needed. In densely populated areas such as cities, where land is expensive, the cost for the land might be prohibitive. And in some parts of the country, such as the Southeast, water hyacinths could clog the waterways. The chief institutional obstacle to such apparently benign systems arises from local, state, and federal regulations. Generally, members of the boards that are involved with wastewater plants are civil and sanitary engineers, accustomed to the familiar mechanized plants. The regulations they promulgate favor mechanized systems—big ones.

THE EARTH'S HEAT

Geothermal energy occurs as a result of tectonic activity and radioactive decay deep within the Earth. According to the United States Geological Survey, "hotspots" capable of yielding the equivalent of 1.2 trillion barrels of oil lie untapped in the western United States and parts of the Gulf Coast region. The most extensive type of hotspot is in the form of hot dry rock, but this is the most difficult type to harness, and the Department of Energy sees such hotspots as contributing little to domestic energy supplies for some time to come.

The next most plentiful hotspots are geopressurized hydrothermal resources—essentially, hot water aquifers containing dissolved methane. There could be extremely large reserves, especially along the Gulf of Mexico. Data on these aquifers comes from nearby petroleum operations; more information, now being gathered, on the number, location, size, permeability, and methane content of these resources is needed before we can know if they will prove economically feasible. Less plentiful—but already in use in many parts of the world—are convective hydrothermal resources, which are systems of hot water and steam heated by relatively shallow masses of hot rock and trapped in fractured or porous rock sediments overlain by impermeable surface layers.

The only active dry-steam reservoir in the U.S. is the system called The Geysers in northern California, where water is pumped into the ground and dry steam emerges to drive electrical turbines. Capacity at The Geysers is now 660 megawatts, with an additional 1400 expected by 1987. Its chief problem is the emission of hydrogen sulfide: the average concentration is 200 parts per million, and the emissions often exceed California air quality standards. A control technology is presently being tested.

Each type of geothermal resource calls for its own technology, and the federal government has been active in promoting the development of these technologies, aiding such efforts as a study of the geothermal wood waste plant near Susanville, California, and the wellhead generators in Hawaii. A government task force has concluded that with an expanded federal program, the United States could develop 20,000–30,000 megawatts of geothermally generated electricity by 1985, and as much as 100,000 by 1990. Full utilization of geothermal energy requires only straightforward engineering progress, not revolutionary scientific advances. Its environmental problems seem relatively easy to solve, and current costs for steam geothermal generation are competitive—about 5.6 cents per kilowatt hour, compared to about 7 cents for coal-fired plants.

Yet there are other, perhaps more interesting, applications of the heat below the Earth's crust. Today, the equivalent of over 7,000 thermal megawatts of geothermal resources are used worldwide for space heating and cooling, agriculture, aquaculture, and industrial processes. More than 50 percent of the space heating requirements of Iceland are provided by geothermal resources. The Soviets use 5,000 thermal megawatts of geothermal heat in agriculture. Each application takes a different temperature range: agriculture is the lowest and industrial process heat is the highest. Such applications of geothermal resources cost about the same as the corresponding fossil fuel uses. As fossil fuel prices rise, geothermal applications will be correspondingly cheaper. Such direct use pays for itself in five to ten years, and this time span can be reduced by what is called "cascading." In Otake, Japan, geothermal fluids cascaded: used first for electrical power production, then space heating, cooking, and bathing—with each subsequent use requiring lower temperatures.

Geothermal energy can, to a degree, be decentralized, especially in direct-use applications. Obviously, it depends on where the hotspots are, and finding them is a chancy business, like drilling for oil, but the heat can economically be transported up to forty miles. Table 5 shows the extent of

Table 5. Worldwide Direct Use of Geothermal Energy

County	Space Heating/ Cooling (MWt)	Agriculture/ Aquaculture (MWt)	Industrial Processes (MWt)*
Iceland	680	40	50
New Zealand	50	10	150
Japan	10	30	5
USSR	120	5,100	—
Hungary	300	370	—
Italy	50	5	20
France	10	—	—
Others	10	10	5
USA	75	5	5
Total	1,245	5,570	235

*Megawatts: Thermal

worldwide direct uses of geothermal energy. In the future, the appealing economics of such direct uses will become a major growth factor for this energy source.

Federal research efforts in geothermal energy development may be reduced in the coming years. The Reagan energy budget cuts geothermal programs by about one-third; the budget cuts affect such programs as the international research effort between the United States, Japan, and Germany to find technologies to exploit "hot, dry rocks" in New Mexico. On the other hand, business interest in geothermal exploitation has grown—for example, from the famed Hunt family of Dallas, Texas. The Interior Department reported in December, 1980, that the Hunt family (including ancillary companies and trusts) controlled 468,000 acres of public land geothermal leases—almost 20 percent of the nation's total public holdings.

As conventional energy costs increase, geothermal energy development will increase. New technologies will render it a viable economic source in the future. Many current schemes have been proposed to exploit the essentially limitless heat energy of the earth—including the detonation of fusion bombs in underground salt domes to simulate geothermal heat—and we can expect to see the commercial development of many of these approaches in the future.

WIND POWER

An enormous quantity of solar energy—estimated at 170 trillion kilowatts—is intercepted by our planet. When the energy strikes the upper atmosphere of the Earth, the phenomenon called wind results. Throughout history wind energy has been used principally for sailing ships and for pumping water by means of windmills.

The popularization of windmills came several centuries after the fall of Rome and opened a new chapter in the history of agriculture and small industry. One reason for the delay was Rome's heavy dependence on slaves. (Emperor Vespasian was opposed to mills because he was afraid they would create unemployment.) But by the eleventh century, almost every European village with access to a running stream had a waterwheel for grinding flour. Progress continued, and a deed written by a Norman in 1180 describes the earliest type of horizontal axis windmill; such devices were common in North Europe by the end of the thirteenth century. By the eighteenth century, the English had developed sophisticated mills with small fantails—wind-powered devices to help swivel the big wooden windmill blades into the wind.

The wind has been used for centuries to propel sailing ships; the last great windjammer, the German *Preussen,* built in 1902, carried 8000 tons of cargo at speeds of 7–10 knots. Today, there is renewed interest in developing wind-powered merchant vessels. In 1980, the Japanese wind-powered oil tanker *Shin-Aitoku Maru* underwent sea trials, and her owners predict that the use of the special plastic sails will save them fifty thousand a year in deferred fuel costs. Other designs for modern wind vessels incorporate computer-controlled sails and other modern accoutrements. Lloyd Bergeson, ship designer and contributing developer of the *Polaris* submarine, believes that a new era of sail-power could "ultimately fulfill 50 percent—or even 75 percent—of all ocean transport requirements." Even a 15 percent reduction in the world's petroleum devoted to the merchant fleet would result in $7 billion yearly savings.

On land, the winds have been a great source of motive power for the industrial revolution. In America, the use of windmills in 1850 represented 1.4 billion horsepower-hours of work—to get the same results from coal would have required buring 11 million tons in that year. In the late nineteenth century, European windmills contributed the equivalent of a billion kilowatt-hours of electricity, in mechanical power. In the 1890s, an enterprising Danish scientist, Poul La Cour, used a wind generator to produce electricity to electrolyze hydrogen and oxygen from water. He used the gases to illuminate the high school at Askov, Denmark.

Until the oil embargo, few large experiments in wind power were attempted, with the exception of the construction of a 1.2-megawatt wind generator at Grandpa's Knob, Vermont, during the Second World War. This large machine (2 blades, 175 feet diameter) was successfully tested by a Vermont utility, but a wartime shortage of parts forced a shutdown for two years. In 1945, a blade failure occurred, and the wind power experiment was terminated. Today, a number of new techniques are being explored to exploit the power of the winds.

Estimates of the potential for U.S. energy production from wind turbines—which are basically propeller designs hooked up to a relatively simple turbine—are wondrous. Theoretically, the wind alone could produce the same amount of electricity in the United States as we produce today from all other sources. Put another way, two hundred thousand wind turbines of a 1 megawatt size could produce the equivalent of 9 million barrels of oil a day—more than we presently import. It would take 67,000 square miles of very carefully selected land—about 2 percent of the nation's land area.

What then are we waiting for?

There are a few technical matters. The key to harnessing the wind is, of course, to understand that it does not blow all the time, being a random and poetic sort of energy source. But recent technical tests have shown that wind power is feasible with turbines up to several hundred feet in diameter. The main drawbacks are economic, and the economic questions relate to performance, reliability, and service life.

The illustration in figure 16 shows a leading commercial design: the 2.5-megawatt "MOD 2" wind generator system, designed by the Boeing Company for the federal government. This horizontal axis design can accommodate either two or three blades; the wind rotor is mounted on top of a tall tower to take advantage of the wind speeds, which increase with height above ground level.

An experimental turbine, operating in Clayton, New Mexico, under the auspices of NASA has shed light on all of these considerations. First, it has achieved the predicted initial levels of power. But actual energy production during the year was half of what had been hoped for. This, the engineers think, was due to a number of problems that would not exist in a production model. Components, chiefly the blades,

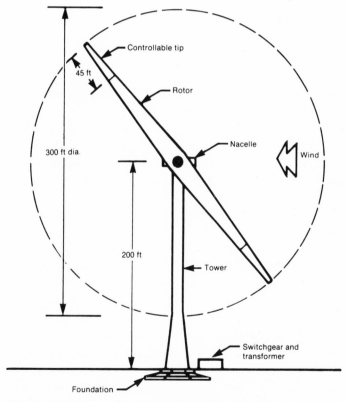

Figure 16. Horizontal Axis "MOD 2" Wind Turbine

failed, and a number of what now seem to have been unnecessary safety shutdowns took place. Design and testing can solve the component problem: new materials are being tested for the blades—from aluminum and wood to composite materials—using a new adhesive that might previously have been thought of as glue but is now a promising way of making super-strong objects by binding various materials chemically. New designs are taking care of the fatigue problem—not only of the rotor, but also of the hub and towers.

At this stage, it is not a matter of how to solve the technical problems but simply when they will be solved. Huge wind turbines, up to three hundred feet in diameter, could be in production by 1982. Smaller ones could be produced commercially during 1981. Both publicly and privately funded prototypes are well along in design and construction.

There are, as we mentioned, ample wind resources. The difficulty is finding them; this takes a great deal of what has come to be known rather charmingly as wind prospecting. Can you think of some poetic friends who would like to be wind prospectors, ambling about the landscape like latter-day Thoreaus and measuring windspeeds? It is not farfetched. Indeed, it is exactly what needs to be done in order to find a place that has winds of fourteen miles an hour or more as an annual mean. Then someone has to measure the wind hourly over a period of years. This is true whether the plan is for a an installation of huge wind turbines in clusters, called windfarms, or whether the plan is for a single, small installation. Prospecting is now being carried out in a number of places, notably Hawaii and California.

Figure 17 shows estimates of the wind resource base, as measured by Battelle Laboratories for the U.S. Department of Energy. As the map shows, there are vast areas of the United States that have wind speeds that can support commercial windpower, including areas of the Northeast, the Great Plains, the Western states, and the Appalachian mountains.

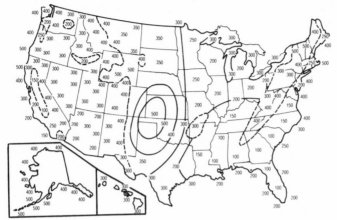

Figure 17. Average Annual Wind Power (Watts/m²) At 50 Meters

Moving from the prospecting stage to the actual installation of a wind power facility that could feed power into a utility grid is an awesome undertaking. One company in particular has become a major force in the introduction of wind technology for commercial applications.

WINDFARMS

Windfarms, Ltd., of San Francisco was formed in 1978 to develop commercial wind energy projects for utility companies. This company identifies utilities that have high oil costs; then they find good wind resources in the service area of the utility and contract to sell energy from wind projects to the utility. The staff of twenty includes meterologists, engineers, systems analysts, land acquisition people, managers, and financiers skilled in assembling complex energy projects. Windfarms does not manufacture its wind turbines but instead purchases equipment suited for individual projects.

In 1979, Windfarms entered into a twenty-five–year agreement to supply electric power to Hawaiian Electric

Company from the world's largest renewable energy project, serving 800,000 customers on the Island of Oahu (including the city of Honolulu). Windfarms has agreed to build a "wind farm" with an 80-megawatt capacity and to deliver the energy over a 19.2-mile dedicated transmission line. The project will be built on the northeast corner of Oahu at Kahuku Point. The $350,000,000 project will consist of twenty 4-megawatt horizontal axis wind machines. According to Windfarms' president, Wayne van Dyck, "When completed in 1984, the project will produce approximately 10 percent of the energy used by the residents of Oahu, and Honolulu and will save about one million barrels of oil annually."

The turbines planned for the Kahuku Point wind energy project will consist of two fiberglass blades with a rotor diameter of 265 feet driving a 4-megawatt generator through a step-up transmission. The 265-foot diameter wind machines will be controlled by microprocessors constantly monitoring the varying wind speeds and adjusting the prop pitch to maintain constant generator speed.

According to van Dyck, Windfarms is also developing projects on the islands of Maui and Hawaii. On the Island of Hawaii, eight 500-kilowatt vertical axis machines will be erected on the side of Mauna Loa at approximately 4,000 feet elevation. These machines will be interconnected with the Hawaii Electric and Light grid. On Maui the company plans a 10-megawatt project consisting of twelve 800-kilowatt horizontal axis machines." A new project proposed for California will use eighty 4 megawatt machines and will cost $700 million.

ENVIRONMENTAL CONSIDERATIONS

As with any energy system, environmental considerations must be taken into account. It has been noted that wind generators might interfere with television, radio, and microwave transmissions. In the latter case, these are narrow-zone waves, and it would be easy to place the machine outside the wave pattern. Big metal-bladed turbines might well effect TV and radio signals; other materials lessen this effect.

Large wind machines do make a certain amount of noise. In Boone, North Carolina, where the federal government has an experimental installation, local citizens have complained of the noise, and engineers are working to change the configuration of rotor and tower to eliminate the noise. Indeed, noise is not a particularly difficult problem to overcome; it is also not likely that very large machines would be placed in population centers. Windfarms would necessarily be located in relatively remote, unpopulated areas.

Land use is a serious consideration inasmuch as windfarms might compete, for example, with timbering. Many of the best sites for windfarms appear to be on national lands: a windfarm will not bother cattle, but quite obviously it would make mincemeat of a wilderness area. The Department of the Interior might well take the lead in examining these lands and considering the ever-controversial question of multiple uses. There are, however, other problems, such as wind rights. Upwind obstructions like buildings or other wind turbines could impede airflow. There is no law on this subject; wind system owners might presently have to buy easements in surrounding land—particularly in suburban and urban situations. Perhaps the greatest impediment to large-scale windfarms is in the area of licensing and permits. From state to state, the requirements for environmental impact statements varies greatly on such matters as land use and building and zoning permits.

Even with small-scale systems, institutional limitations exist: zoning restrictions limiting height, setback, competing energy resource, and aesthetic considerations could severely limit the use of residential and commercial wind systems. Building, safety, and housing codes have not been developed from the standpoint of wind power. The user of the small-scale system, moreover, is likely to have conflicts with the utility. The user who wishes to connect to the grid may be required to install and maintain control and protective devices to insure safe operation parallel to the utility's generation facilities, and these plus the high cost of insurance against acts of God and vandalism weigh heavily against the economics of small-scale wind systems. Yet, as the technology develops, these costs should come down. At this point, the industry itself remains small-scale and undercapitalized. In a sense, it is similar to photovoltaics: there is a potential market if the price could be reduced, and the price could be lowered by mass production—if the market were ready.

The federal government has already begun to intervene in this capital-market equation by providing, for example, tax credits and faster depreciation allowances. These could be augmented in a number of ways, including government purchase of large wind systems. Dr. James Lerner of the California Energy Commission is perhaps one of the most careful thinkers today on this subject, and the Energy and Defense Project report drew heavily on his insights. We summarize some of his thoughts here.

The near term potential market for wind generators is in areas served by utilities that have large amounts of oil-fired capacity and good wind resources, such as Hawaii, California, and the Northeastern United States. The utility market should become viable by about 1983–1985, while the residential market may become viable by the mid to late 1980s. These dates depend on future oil price rises and availability, government incentives, technical and institutional considerations, capital availability, and the rapidity with which a viable wind system manufacturing and distribution system can be developed. Demonstration of credible performance and proven system reliability and removal of the major institutional barriers to siting wind systems will have a significant impact on the development of this market. Federal and state programs and policies could also have a major impact on the development of windfarms.

The emergence of the various markets for wind systems is a complex process that is only now beginning to be quantified. For example, the fact that utilities will accept a thirty-year-pay-back period has a great deal to do with the early development of that market. The same may be true of residential users who normally accept twenty-to-thirty-year mortgages when purchasing a residence. Generally, business investors tend to look for a high rate of return on investments with pay back periods of five years or less. This latter characteristic of business investors may tend to delay certain industrial, commercial, or agricultural applications of wind systems until the late 1980s, well after the date the technology is completely proven.

Utility windfarms employing clusters of medium- or large-scale generators could be considered a "centralized" application, while residential and other applications employing single units or clusters of several units—in unit sizes from several kilowatts to large-scale systems in the megawatt class —could be considered a "decentralized" application. But the "centralized" category does not quite fit the traditional definition: since the windfarm covers a much larger land area than a conventional power plant of comparable size (the same land area as covered by a conventional plant of ten times the capacity), this "centralized" application can be considered less centralized than conventional large central station fossil or nuclear plants. Over thirty times as many windfarms would be required to satisfy the U.S. electrical demand—compared to large centralized fossil or nuclear plants. In terms of vulnerability, the windfarms would be less vulnerable to attack or sabotage owing to the fact that there would be thirty times as many plants, and the individual units would be dispersed over a larger area.

There are numerous non-grid applications of wind systems for remote or isolated energy consumers. Since many of these isolated users currently pay more than 10 cents per kilowatt hour for electrical energy, this is a ready market for

wind generators. Non-grid-interconnected applications include telecommunications, isolated utilities, offshore oil and
gas platforms, onshore oil and gas pipelines, defense installation, navigational aids, rural residences, and farms—all total-
ing perhaps more than two million wind systems of various
sizes. However, each application is constrained by the need
to provide energy storage systems. Many types of energy
storage systems have been proposed for use with wind power
systems; however, many of these options, such as hydrogen,
thermal, flywheel, and compressed air storage systems, are in
the conceptual or early experimental stage, and their associated energy storage costs are not well defined.

The key to maximum utilization of wind energy and maximum oil savings rests with the use of grid-integrated wind
systems. Without some form of energy storage system, these
generators will probably be limited to 10 to 20 percent of the
electric supply as fuel savers. One very promising approach
for avoiding the temporal, seasonal, and geographic limitations of more extensive deployment of wind technology is
through the use of hybrid or integrated systems of renewable
technologies that incorporate wind generators with other
renewables, such as wind and hydro, wind and biomass, or
wind, solar, and hydro. The wind-hydro integration appears
to be particularly attractive. Two alternates are possible: the
use of wind for peaking, by conserving water during periods
when the wind is blowing for controlled release later through
the hydroelectric turbines during peak demand periods, and
the use of wind to pump water to a higher reservoir during
base demand time and release of the water to generate
power during peak demand.

Community self-sufficiency becomes a distinct possibility
for certain locales that possess the "correct" blend of renewable energy resources. Wind, or any single solar or renewable
resource, taken alone, cannot provide self-sufficiency; however, a hybrid combination of two or three renewables does
offer this possibility. A community must first identify its re-

newable energy resource potential, then consider scaling and compatibility of various renewable technologies, examine engineering feasibility, consider load management, costs, financing, institutional and legal factors, and community access and acceptance. Wind energy used in conjunction with hydro, biomass, geothermal, and possibly solar photovoltaic (when this alternative becomes less expensive) offers some promise for community self-sufficiency.

In New Hampshire, a feasibility study is underway to investigate the use of wind in conjunction with an upgraded 750 kilowatt hydro site to provide for reliable year-round cost-effective operation. In the absence of wind, the hydro facility can operate only for about eight months of the year. The wind turbine would probably be used in the pumped storage mode to stabilize peak energy requirements needed during the winter nights. It will be hooked into the utility for backup in the event of a drastic lack of water or wind.

In such applications, wind power may prove to be a major energy contender in the future. Certainly, in the right locations, such as the windy areas or the high mountain ranges of Hawaii, wind power shows great promise. Of all the solar energy technologies, wind machines appear to have the best economics, surpassed only by the near-term applications of biomass resources.

BUILDING FOR THE FUTURE

In chapters 9 and 10 we explored a number of energy sources either derived from the sun, or related to solar energy conversion. In some quarters, we are told that the solar age has nearly arrived. Denis Hayes, director of the federally sponsored Solar Energy Research Institute, says that "Forty percent of our energy could come from solar energy by the year 2000 if we make some dramatic moves *now.*" Former President Carter formally endorsed the goal of 20 percent of the nation's energy coming from solar and renewable sources of energy by the year 2000. Yet, in the last days of the Carter administration, his handpicked Council on Environmental Quality said that there was "no coordinated strategy" for achieving this 20 percent goal.

In our review of the various solar and renewable technologies, we recognize the potential of many solar-derived techniques, but we question the logic of some of them. For example, solar water and space heating technologies appear to be largely inadequate for vast areas of the country and highly expensive for those areas where they are compatible. A far better alternative is the use of sound conservation approaches and the implementation of "passive" solar design. OTEC technologies (Ocean Thermal Energy Conversion) are grossly expensive and are not likely to be economically competitive, except in isolated locations. On the other hand, a number of solar-derived technologies seem to be capable of providing reliable power, on a large scale and when sited properly. Ranging from wind power to various biomass approaches, these techniques are worth further funding and serious attention.

There are inherent problems in the large-scale application of certain biomass technologies. Over the past few decades, a number of attempts have been made to create "energy farms"—growing algae, kelp, or various plants under artificial conditions to achieve high yields for food and energy. In many cases, the energy expended on the experiments has been greater than the yields. Ecologist Howard T. Odum noted on one set of experiments on algae that "as an appreciable yield develops, energy laws require a decline. . . . In the laboratory tests, [ecological supports and minerals] were supplied by the fossil fuel culture through thousands of dollars spent annually on laboratory equipment and services to keep a small number of algae in net yields." We should be careful not to assign great priority to such attempts to "beat" photosynthesis through fossil subsidies. On the other hand, we must not lose sight of the necessity of using our conventional energy sources as a necessary bridge to newer and less centralized ones. As we have seen in this chapter, coal conversion to methanol will help pave the way to wider adoption of renewable-based alcohol fuels in the future.

Finally, the role of government in determining energy systems is vast. Battelle Laboratories have calculated that the the federal government has provided subsidies of about $252 billion to the conventional energy industries over the past few decades. These subsidies include, for example, depletion allowances, research funds, tax incentives. Table 6 shows the distribution of government subsidies.

It is encouraging to see that federal government attitudes on the subsidy issue are changing, but we take issue with the priorities of the Reagan administration. As we have noted, government subsidies for synthetic fuels development did not disappear—they have re-emerged, in large measure, through congressional intervention. Likewise, funding of new programs for centralized nuclear programs has increased dramatically with the Reagan budget, and these programs comprise 53 percent of the Department of Energy budget. Conservation programs, which have suffered the most severe cuts, account for only 6 percent of the department's budget. A 1981 proposal by a group of environmental organizations suggested a budget that would reduce the nation's 1982 deficit by $1.7 billion more than the administration's budget. This would be accomplished by reducing and eliminating the numerous federal subsidies to energy-intensive enterprises, such as centralized energy projects, continuing the Interstate Highway System, and others. In addition, the environmental groups proposed user fees for harbors and highways to increase federal revenues.

We do not believe that the federal government will necessarily be instrumental in bringing about the changes we suggest in the U.S. system, but until the era of subsidies for energy centralization ends, there is little chance for most of the small-scale technologies that we advocate to compete fairly in the marketplace.

As we saw in the introduction to Part Two, the Aspen Institute conference identified a number of new and innovative technologies that can play a major role in making the United States less dependent on foreign energy sources. These technologies have been discussed here and, for the most part, they are available today. The Aspen conference cautioned that many of these (especially the solar ones) are "highly capital, materials and energy-intensive" initially, so they need to be discounted over twenty to thirty years.

We believe that these technologies will play an increasing role, but national policies must recognize that free opportunities—for research dollars as well as market opportunities—must be established and protected. The stakes are high, and the rewards are great. By adopting these policies, and encouraging a new era of decentralization and self-sufficiency, we can take a major step away from the present course toward nuclear holocaust.

Table 6. An Estimate of the Cost Incentives Used to Stimulate Energy Production
(in Billions of 1978 Dollars)

	Nuclear	Hydro	Coal	Oil	Gas	Electricity	Percent of Total	Total Incentives
Taxation	—	2.0	4.74	55.48	14.92	38.83	115.97	46.0
Disbursements	—	—	—	1.30	—	—	1.30	0.5
Requirements	1.7	0.04	0.80	57.49	-0.80	—	59.23	23.5
Traditional Services	—	—	2.57	6.92	—	0.52	10.01	4.0
Nontraditional Services	17.2	—	3.55	1.88	0.30	—	22.93	9.1
Market Activity	2.1	14.86	—	0.02	0.50	0.15	25.17	42.80
Totals	21.0	16.90	11.68	123.57	14.57	64.52	252.24	100
Percent of Total Incentives	8.3	6.7	4.6	49.0	5.8	25.6		100

Epilogue: Toward Defense

IT IS an axiom of sports that the best defense is a good offense. The world's military planners seem to have taken this too much to heart. The United States and most other nations have no real defense against nuclear war. As former National Security Council official Jan Lodal comments, "Strategic defense is completely absent from U.S. nuclear strategy. There is no reference to defense in published statements of strategy, and the United States maintains no effective defense against Soviet attack. Our air defenses are limited, and the single Safeguard site permitted by the ABM treaty is mothballed. Nor do we have a significant civil defense program." Instead, we have put our faith in "Mutually Assured Destruction"—MAD—and we have acquired thereby a Department of Offense.

While the Reagan administration has moved to increase defense spending by $40 billion in 1981 and 1982, there has been no related commitment to engaging in the Strategic Arms Limitation Talks (SALT), stalled by President Carter in the wake of the Soviet invasion of Afghanistan. Both Russia and the United States are moving toward military confrontation—as we have seen, the Soviets have greatly increased military spending and are now launching several new weapons systems. Sidney Drell of Stanford University says that "arms control is an important part of our national security. Thus far we have had no effective controls on offensive nuclear weaponry, and it is clear that each step forward in the arms race to more and improved weapons has *lessened* our security. If we are to reverse this trend it will be necessary to understand the arms control impact of new weapons before making a decision to deploy them."

Clearly, we must take the issue of defense and arms control seriously. The nation's mood about impending nuclear war has shifted toward the fear of the 1950s and 1960s, and William Kincade argues that we will have to accustom ourselves to the insecurity experienced by the Europeans since World War II. They have, he says, lived with this insecurity, "with their survival dependent on decisions made in distant capitals, and possessing no immunity from nuclear weapons effects. Increasing accuracy, shorter times to target, reduced decision or response time, larger numbers of more versatile weapons and lengthening lists of feasible targets are creating circumstances for Americans similar to those Europeans have experienced for twenty-five years: life at the epicenter of a potential nuclear battlefield where the likelihood of intentional or accidental war is remote, but never quite remote enough." As matters now progress he notes, "the likely effect of the technological developments foreseeable in the next decade or two will be not a rising sense of security, but growing fear and doubt."

The art of preventing war has to do with stabilizing our society—of doing all the things necessary to strengthen our civilization in the face of the apocalyptic alternative: the horsemen of death, war, famine, and judgment. In this book, we have argued that numerous approaches and technologies are available *now* to speedily relieve our nation of the perilous dependence on strategic materials and energy.

But we need to address the realities of a world in which the Superpowers are investing more than $100 million *each day* to expand the arsenals of atomic might. With more than fifty thousand individual armed nuclear weapons in place, the atomic powers have amassed a destructive force a million times the power of the Hiroshima bomb.

Dr. Vannevar Bush, chairman of the National Defense Research Committee during the Second World War, wrote in 1949 that the arms race would be determined by "which system of government and industry is the more efficient and has the better staying power." He elaborated,

> To win the race we must have a healthy people. We must raise our standard of living so that more of our population may perform well. We must learn to make our industrial machine operate smoothly and avoid the interruptions because of quarrels over the division of the product. We must learn to avoid inflation and depression. We must somehow produce governmental machinery that will operate efficiently for its intended purposes, so that the selfish interests of groups or sections cannot drain away our energies. We must establish justice and good will among our people and among the races that make up our population, so that our progress will not be halted by internal friction. We may not accomplish all these things, but we should accomplish most of them if all our citizens realize with full clarity that the alternative is someday to enter an atomic war on the losing side.

In the three decades since Bush wrote those words in *Modern Arms and Free Men,* his goals have not, unhappily, been met. The world is less stable now than at any time since the last great war. Inflation is rampant in the world's economies, especially threatening the countries that do not produce oil. Other signals of instability, such as unemployment, are present as well; unemployment rates are rising in developed and poor countries alike. Simultaneously, there appears to be an almost direct correlation between military spending and declining economies. Military budgets in countries around the world inhibit the necessary development of capital for economic growth through their drain on research funds, productivity, and talent.

As the world's military budgets increase, and more and

more attention is paid to aggression as the preferred answer to disputes, the security of all nations is threatened. What we suggest as a national resource policy for the United States is also a key international objective. In a recent report, *North-South: A Program for Survival,* published by the Independent Commission on International Development Issues, its chairman, Willy Brandt, speaks plainly: "If reduced to a single denominator, this report deals with peace. War is often thought of in terms of military conflict, or even annihilation. But there is a growing awareness that an equal danger might be chaos—as a result of mass hunger, economic disaster, environmental catastrophes, and terrorism. So we should think not only of reducing the traditional threats to peace, but also of the need for change from chaos to order."

The Brandt report calls for a strong program of international development to increase the availability of food and basic resources to the poor nations of the world. On energy, the commission urges alternative development in Third World countries and asks the industrialized nations to set ambitious targets for energy conservation and to minimize oil consumption.

DECENTRALIZATION AND WARFARE

The world's nuclear arsenals have been built up by nations that have consistently centralized their technological establishments and devoted great wealth to this enterprise. Accumulating such force is beyond the scope of anything but large and wealthy nations. In *The Breakdown of Nations,* published in 1957, Leopold Kohr argued that "whenever a nation becomes large enough to accumulate the critical mass of power, it will in the end accumulate it. And when it has acquired it, it will become an aggressor, its previous record and intentions to the contrary notwithstanding." Kohr concluded that the forces of centralization inexorably lead to "social barbarism." However, "in a small society, the critical quantity of power can only rarely accumulate since, in the absence of great numerical weight, the cohesive force of the group is easily immobilized by the self-balancing centrifugal trends represented by the numerous competitive pursuits of its individuals."

We are witnessing, in the waning years of the twentieth century, the thrust of military centralization, which now permeates all of the industrialized nations. Yet even the best military minds, bent on assembling vast and precise arsenals, are not unaware of the virtues of decentralization.

Historically, decentralizing defensive weapons has been considered essential. The triad of land-based missiles, strategic bombers, and submarine carried missiles rests on this principle. The new MX missile system, proposed for deployment in Utah in a huge array of underground railroads that carry missiles back and forth to multiple shelters, is designed precisely to disperse potential targets. (It is also a little ironic to note that the MX system is to be fueled by alternative energy sources such as fuel cells and solar energy, decreasing the system's vulnerability to cut-offs of fossil fuel supplies and to the effects of electromagnetic pulse.)

Some military planners think that even the MX land-based system is too centralized. A defense consultant, Richard Garwin, has proposed a water-based submarine version of the MX in which shallow water submarines would carry two MX missiles apiece. It would be less expensive and a more reliable protection for the missiles.

It is nice to have such a serious debate in behalf of missiles. We would like to see a similar debate in behalf of our population—particularly with regard to resource contingency planning. We remain extremely vulnerable—with or without a nuclear attack. We believe that decentralizing and strengthening our energy systems will not only serve local populations better but will act as a primary deterrent to nuclear war, by reducing our dependence on foreign sources of fuel and by increasing the resiliency of our energy system in the event of war.

It is our contention that our present dependence on foreign resources and the high degree of centralization (and the resulting inefficiences of centralization) render this nation dangerously unstable and vulnerable. The idea of inviting trouble because of weakness or disarray is nothing new. During the debates about the formation of the Union, John Jay argued that the American people wisely "consider union and a good national government as necessary to put and keep them in *such a situation* as, instead of inviting war, will tend to repress and discourage it. That situation consists in the best possible state of defense, and necessarily depends on the government, the arms, and the resources of the country."

What constitutes an invitation to war? In considering nations, one can find some useful analogies from the realm of nature. Wildlife biologists have found that even the most lethal of predators rarely attack a healthy member of a prey species—instead, they pick out the weak, the sick, the old.

On the other hand, among such social animals as wolves, there is a definite hierarchy, at the top of which is what is called the alpha male, who dominates the pack and does so with virtually no challenge. In his classic study, *The Wolf,* biologist David Mech pointed out that the "dominance shown by the alpha animals . . . can be described as a kind of forceful initiative. When a situation does not require initiative, dominance may not be shown; for example, when a pack is resting. However, when food, favored space, mates, strange wolves, or other stimuli are present, initiative can be seen in the action of dominant animals."

And in evolutionary theory, it is the unstable species that faces either a rapid evolutionary change or extinction. Indeed, the two processes are hard to separate. We, whose most ancient progenitors were mammals hiding from the dinosaurs 300 million years ago, have the capacity to change our course, to direct it toward goals that are, in the biologist's terms, adaptive. It is time to let some of the dinosaurian elements of our civilization die so that we may evolve into something different—a nation that is stable, flexible, and strong. The tools are at hand and more could rapidly become available.

Appendix: Selection of Alternative Fuels and Electric Power Sources

The technical advisory committee for the Energy and Defense Project evaluated a number of the alternative resources available at the regional and local level. We found that increasing the energy self-sufficiency of communities and regions can be accomplished by integrating a combination of available dispersed and renewable energy technologies. Two types of technologies were evaluated: fuels and electricity. The matrices reproduced below illustrate the properties of the major technologies; they are valuable guides to the selection of energy resources, to the economics of various decentralized energy sources, and to many other characteristics of the technologies. May they serve as a focal point for the energy development of the future.

EXPLANATION OF MATRIX CATEGORIES
FOR FIGURES 18 AND 19

Categories in the matrices are judged from a strategic perspective, based on criteria of availability (local and regional), current and projected costs, and overall flexibility. The rank is from 10 (best) to 0 (worst). The categories are expressed primarily as Y (yes) or N (no, not applicable). Alternative categories are L (low), M (medium), H (high); in some cases, a range is expressed (L-H), or dual flexibility (Y/N).

Rank is evaluation on a scale of 1–10, with 10 having the highest value, judged from a strategic perspective. In fuels, high ranks designate the suitability of a fuel from a local and regional production and use basis. Flexibility, renewability, ease in production, and other key characteristics affect the ranking. In electricity, the same strategic evaluation applies, with some technologies that are inherently dispersed and commercially available, having high rank (cogeneration, small fossil plants, and so forth). Technologies such as photovoltaics and wind power are renewable and available but are ranked lower because of current low production and high costs; however, from a community and regional perspective, these are important technologies to integrate in emergency and energy planning.

Dispersed describes the local and regional production possibilities for fuels and electric power. For example, gasoline, diesel fuel, crude oil, and synthetic natural gas are all fuels that require considerable capital investment in high technology production facilities; they are most economically made in large bulk quantities (that is, production runs greater than 3,000 tons or 2.7 million kilograms per day). These fuels are therefore best produced in large centralized facilities (not dispersed) and require distribution networks to reach their ultimate consumers. Methanol, ethanol, biogas, and the other fuels listed in the fuels matrix (figure 18) are more easily produced and are thus evaluated as being good potential candidates for dispersed or decentralized supply systems. It is also economical to produce them in smaller lot quantities (less than 1,000 tons or .9 million kilograms per day). On the matrix, all of the technologies for electricity are capable of dispersion with the exception of geothermal and waves, which are site-specific.

Central. In addition to dispersed fuels (or systems) and

electrical technologies, centralized technologies may also apply to many of the same categories. For example, cogeneration systems may occur in central as well as dispersed locations; methanol and ethanol fuels can be either dispersed or centralized.

Renewable (Renewable Feedstocks). In fuels and electricity, renewable characteristics refer to solar, biomass, wind, water, and other renewable technologies. All of the fuels listed can be made from bio-feedstocks, so they are rated as low, medium, or high potential for commercial production. Electrical technologies are characterized on a simple yes/no basis.

Feedstock/Fuel Flexibility. In fuels, high flexibility refers to use of a variety of feedstocks (biomass and fossil origin). In electricity, flexibility is high if different fuel sources can be used for each category of electrical technology.

Grid-Connected and Grid-Independent. Some fuels and electrical technologies may be either grid-connected or local, and not connected. If a fuel is normally distributed through central systems (pipelines and distribution), it is rated (Y) for grid-connected. In electrical systems, all the technologies can be grid-connected, but for grid independence, some are rated (E) for ease in isolated operation. Some systems are more difficult to operate outside the grid. However, all electrical systems *can be* designed for local operation, independent of central grids.

Local Fuels and Feedstocks. These fuels and sources for electrical power are rated (Y/N) for electricity, based on local availability. For fuels, use of locally available feedstocks is rated (L-H) low-high.

Site Limited and Site Dependent. Site dependent fuels require large fuel stocks and capital investment, as opposed to non-site dependent sources such as Low BTU gas, which can be made in a mobile gasifier transported to dispersed locations. Site dependent electrical technologies such as geothermal or wind are not flexible, like cogeneration systems.

Local and Regional Components refers to the availability

of key components and spare parts of technologies that may be found either locally or within the region where the fuel/electrical process is located. As an example, the production of methanol requires a catalyst material usually not available locally. Likewise, small fossil plants require sophisticated components and spare parts that would not be available locally.

Local Maintenance. Some fuels and technologies can be produced and operated using the local/regional labor force. The ratings are based on the likelihood of availability of this expertise.

Capital Intensive refers to the range of installed costs and the strategic material intensity of the fuel processes and technologies. As can be seen, the production and use of dispersed Low BTU gas is one of the highest rated dispersed fuels and technologies.

Short Lead Time refers, in general, to fuel processes and technologies that can be ordered and delivered for energy production within three years. Gasoline and SNG facilities require many years to license and construct, as opposed to Low BTU gas facilities that can be built quickly. Likewise, micro-cogeneration systems can be built quickly, unlike geothermal or solar thermal facilities, which require years.

Mobility refers to fuels in cases where the production facility can be located at the source of the fuel. In electricity, mobility refers to the flexibility of the power plant's location. Some technologies, such as small hydro, are definitely not transferable from specific sites.

Operation and Maintenance Costs are rated H-L. Maintenance is self-explanatory; operations costs also include labor, capital depreciation, feedstocks, and costs of transportation.

Storage (Fuels). Storage capability is rated high if storage facilities are locally available, and if it makes sense to store the fuel. Hydrogen, for example, is rated low because it is difficult and expensive to store for an appreciable length of time.

End Use Flexibility (Fuels). A fuel is considered to have

a high flexibility if many different converters can be adapted to use of the fuel (boilers, turbines, internal combustion engines). Obviously, fuels such as gasoline have high flexibility.

Scale (Fuels). The scale of production for fuels is rated L (large) for production processes that are greater than 3,000 tons/day (2.7 million kg) equivalent, M (medium) for processes operating at 1,000–3,000 tons/day (.9–2.7 million kg) equivalent, and S (small) for processes less than 1,000 tons/ day (.9 million kg) equivalent.

Size Range (Electricity). The size of average technologies and processes is expressed in MW (one megawatt equal 1,000 kw); photovoltaics, for example, are used in an average configuration of panels that are small (a few kilowatts), but can be up to ten MW in power "farms." The same is true for wind generation.

Intermittent (Electricity) refers to technologies that may be seasonal in nature, such as small hydro, or operate only during sunlight (solar systems), thereby requiring energy storage for baseload operation.

Costs. Fuel costs are expressed in current dollars/million BTUs for fuels at the refinery gate or production site. These costs include amortization of capital investments. Electricity costs are expressed in capital costs per installed kilowatt of capacity ($/kw). These costs represent current costs, not estimates of future costs of the technologies.

EXPLANATION OF FUELS MATRIX

Gasoline is a premium fuel that can be used in stationary or mobile applications. However, in current refining practices, lower quality heavy crude oil will not produce as much gasoline as lighter crudes (previously in greater abundance). It can also be produced from methanol (via a Mobil Oil Company process), however, allowing significant resource flexibility (biomass, coal, natural gas, shale, heavy crudes, tar sands, and so forth)

Diesel Oil is a middle distillate. This category (on the

DERIVED FUELS	Rank	Dispersed	Central	Renewable Feedstock	Feedstock Flexibility	Grid-Connected	Grid-Independent	Operation and Maintenance Costs	Local Feedstock	Site Dependent	Regional Components	Local Maintenance	Capital Intensive	Short Lead Time	Mobility	Storage	End Use Flexibility	Scale	$/Million BTUs
Gasoline	3	N	Y	L	H	Y	N	H	L/H	Y	N	N	Y	N	N	H	H	L	5.50 to 7.50
Diesel	6	N	Y	L	H	Y	N	H	L/H	Y	N	N	Y	N	N	H	H	L	5.50 to 8.00
Crude Coal	5																		
Crude Shale	9	N	Y	L	M	Y	N	L/H	H	Y	Y/N	Y/N	Y	Y/N	N	H	L	M/L	5.00 to 7.00
Methanol	8	Y	Y	M	H	Y	Y	M	H	Y	N	N	Y	N	Y/N	H	H	S/L	6.00 to 8.00
Ethanol	10	Y	Y	H	M	Y	Y	H	H	Y	Y	Y	Y	Y	N	H	H	S	8.00 to 12.0
Low BTU Gas	9	Y	N	H	H	N	Y	M	L/H	N	Y	Y	N	Y	Y	L	M	S/M	4.00 to 5.00
Med BTU Gas	9	Y	Y/N	H	H	Y	Y	H	L/H	N	Y	Y	Y	Y	Y	L	M	S/L	4.50 to 5.50
Biogas	7	Y	N	H	M	N	Y	L	H	Y	Y	Y	Y	Y	N	L	M	S	4.50 to 8.00
SNG	7	N	Y	L	H	Y	N	M	L/H	Y	N	N	Y	N	Y/N	M	M	L	5.50 to 8.50
Hydrogen	1	Y	Y	H	H	Y	Y	H	L/H	Y	Y/N	Y/N	Y	N	N	L	H	S/L	7.00 to 50.0
Biomass Oils & Lubricants	6	Y	N	H	L	N	Y	L	H	N	Y	Y	N	Y	Y	H	H	S	8.00 to 12.0

Figure 18. Fuel Sources Matrix

matrix) includes all middle distillates, from aviation fuel to kerosene. It is a more difficult fuel to produce from feedstocks other than crude oil and is not as versatile a fuel, from a strategic standpoint.

Crude Oil is a natural oil (as is shale oil), and can be made from coal, tar sands, wood, and other carbonaceous feedstocks. The lower rating (5) relates to coal production, and the higher rating (9) relates to production from new domestic oil resources. Shale oil is a more attractive feedstock than heavy crude oil.

Methanol can be made from all hydrocarbon feedstocks through partial oxidation (gasification). This is an extremely versatile fuel, but catalysts are required to convert producer gas to methanol. This limits production flexibility and reduces local production capabilities. To insure continuous production, an inventory of catalytic materials would be required.

Ethanol can be made from non-renewable resources but is typically made from biomass-derived sugars and starches. This is an immediately available premium fuel that can be used as an independent fuel or blended with other products such as gasoline. The conversion technology is commercially available with locally available components.

Low Btu Gas (LBG) is gas with a maximum heat value of 200 Btu/ft^3, made from hydrocarbon feedstocks. It is made through partial combustion in an air-blown gasifier. This gas can be used in internal combustion engines but cannot operate gas turbines (with current technology). It can substitute for most natural gas uses.

Medium Btu Gas (MBG) is a partially combusted hydrocarbon gas with a heat value of 200–500 Btu/ft^3. It requires pure oxygen in the gasification process, which increases costs and requires additional equipment. This gas can be used as a feedstock for synthetic fuels (methanol, SNG, gasolines, and so forth) and as a fuel in gas turbines and other heat engines, such as boilers.

Biogas is a methane-rich (CH_4) gas with a heat value of 500–700 Btu/ft^3 and can be used as a boiler fuel in gas turbines and other heat engines. It is a substitute for natural gas. It is typically produced by decomposing organic materials that are locally available.

Synthetic Natural Gas (SNG) is a high-heat value gas (1,000 Btu/ft^3) that can be a direct substitute for natural gas in essentially all applications. It is made through cataytic conversion of MBG. Feedstocks include oil, coal, shale oil, and biomass. SNG can also be made by purifying biogas.

Hydrogen (H_2) can be extracted from coal via gasification processes or it can be made by the electrolytic decomposition of water. It is a volatile, high quality fuel that can substitute for natural gas. However, conversion processes are highly energy-intensive, and significant infrastructure problems such as storage and distribution, stand in the way of widespread utilization.

Biomass Oils and Lubricants are vegetable oils that can be derived from locally available oil-producing plants (sunflower, safflower, jojoba, for example). These oils have been used successfully to substitute for diesel fuel, although their strategic significance stems more from their value as lubricants than fuels.

EXPLANATION OF ELECTRICITY MATRIX

Cogeneration is the generation of both electrical or mechanical power and the production of useful heat from the same primary source of fuel. A typical configuration is the use of steam from a fossil-fired boiler to drive a turbine-generator and the subsequent use of the exhaust steam for space or water heating.

Small Fossil Plants are defined as any fossil-fired electric generating plant with an output capacity of less than 250 megawatts. These are primarily steam-driven turbine-generators.

Small Hydro is an electrical generating system with an output capacity of less than 30 megawatts powered by falling, or otherwise moving, water. This source may represent the most thoroughly developed technology included in this dis-

	Rank	Dispersed	Central	Renewable	Fuel Flexibility	Grid-Connected	Grid-Independent	Local Fuel Supply	Site-Limited	Local Components	Local Maintenance	Capital Intensive	Short Lead Time	Mobility	Operation and Maintenance Costs	Size Range (MW)	Intermittent	Cost ($/kw) Capacity
Cogeneration	10	Y	Y	Y/N	Y	Y	E	Y/N	N	N	Y	N	Y	Y	L	0-50	Y/N	500-1500
250MW Small	10	Y	Y	Y	Y	Y	E	N	N	N	Y	Y	Y/N	Y/N	H	0-250	N	500-2000
Fossil Plants																		
Small Hydro	10	Y	N	Y	N	Y	E	Y	Y	Y	Y	N	Y	N	L	0-30	Y/N	600-1000
Wind	7	Y	Y	Y	N	Y	D	Y	Y	Y	Y	Y	Y	Y	L	0-5	Y	1000-2000
Photovoltaics	4	Y	Y	Y	N	Y	D	Y	N	N	Y	Y	Y	Y	L	0-10	Y	10,000 +
Biomass Steam	8	Y	Y	Y	Y	Y	E	Y	Y/N	Y	Y	Y	Y	Y/N	M	.2-50	Y/N	500-1500
Biomass Low																		
BTU Gas	7	Y	Y	Y	Y/N	Y	E	Y	Y/N	Y	Y	N	Y	Y	M	0-5	Y/N	500-1200
Geothermal	10-6	N	Y	N	N	Y	D	Y	Y	Y	Y	Y	N	N	H	5-50	N	700-4000
Fuel Cell	3	Y	Y	Y/N	Y	Y	E	Y/N	N	N	N	Y	N	Y	H	0-5	N	5000 +
Waves	1	N	Y	Y	N	Y	D	Y	Y	N	Y	Y	N	Y	H	?	Y	15,000 +
OTEC	1	Y	Y	Y	N	Y	E	Y	Y	N	Y	Y	N	N	H	?	N	15,000 +
Low Temp																		
Solar Thermal	5	Y	Y	Y	N	Y	E	Y	N	N	N	Y	N	Y/N	L	0-5	N	4000 +
High Temp																		
Solar Thermal	4	Y	Y	Y	N	Y	E	Y	N	N	N	Y	N	Y/N	M	0-10	Y	6000 +
Fossil																		
Gasification	10	Y	Y	N	Y	Y	E	N	N	Y	Y	Y	Y	Y	M	5-50	N	1500-4000

Figure 19. Electrical Generation Sources Matrix

cussion; plants of virtually any size are readily available from commercial vendors.

Wind. Any one of numerous Wind Energy Conversion Systems (WECS) use wind-powered propellors or blades to drive an electric generator. Small systems are commercially available at this time; however, systems in the megawatt range are still in the development and testing state. The size range given here is for individual towers, but much larger outputs might be obtained from wind "farms" of 25 or more units.

Photovoltaic power involves the direct transformation of sunlight into electricity through the excitation of various semiconductor materials. Very small systems are currently in use, but the high cost of high-grade photovoltaic materials currently limits an otherwise wide range of applications.

Biomass Steam is any plant material or waste from plant material that is combusted in a boiler. Such a system may use a Rankine-cycle heat generator to produce electricity or a conventional turbine-generator with fossil-fuel backup.

Biomass Low Btu Gas is produced by partially combusting biomass fuel in a reactor to break the fuel down into its hydrogen and carbon-monoxide components. These two combustible gases may then be burned in boilers or certain combustion engines.

Geothermal-electric power may be produced by utilizing the heat within the earth resulting from either tectonic activity or radioactive decay. The most developed technology uses naturally created steam. These systems, however, are limited by relatively few sites and problems associated with the chemistry of geothermal steam. The United States enjoys extensive "hot dry rock" resources—requiring the injection of water to produce steam—but the required technology is still in the early stages of development.

Fuel Cells are electrochemical devices that chemically combine hydrogen and oxygen to produce electricity and water. The systems have been utilized in specialized applications such as space vehicles, but large-scale applications are in the early development stages.

Waves. The energy of waves may be converted into electricity by the use of wave pumps, pneumatic devices, motion devices, underwater pressure field devices, and facilities powered by the mass transport of water from breaking waves. Very small systems are currently being developed; however, technological obstacles have inibited full-scale development of this source.

Ocean Thermal Energy Conversion (OTEC) produces power from the thermal layer differences between warm surface water and colder deep ocean water. Serious engineering obstacles and a limited number of sites have inhibited development of this source.

Low Temperature Solar Thermal. The most common low temperature solar technology is the solar pond that uses salinity layers in a body of water to absorb and trap solar energy and to convert that heat into electricity through a Rankine-cycle turbine. The technology is in commercial use in several countries and in the testing stage in the United States.

High Temperature Solar Thermal systems use concentrating collectors to focus solar energy on a target. The sunlight can be concentrated sufficiently to produce temperatures up to 2,000°F (1,093.3°C). Water in the receiver is thus boiled to produce steam for turbine-generators. Very small high temperature systems are commercially available but larger systems are in the testing and development stages.

Fossil Gasification systems use an oxygen-blown gasifier to convert fossil fuels such as coal or heavy oil into their carbon monoxide and hydrogen components, which are subsequently used in a combustion system to generate electricity.

References and Additional Reading

CHAPTER 1

The Effects of Nuclear War. Congress of the United States, Office of Technology Assessment. Washington, D.C.: U.S. Government Printing Office, May 1979.

Yost, Charles W. "National Security Revisited." *Bulletin of the Atomic Scientists,* October 1980.

CHAPTER 2

Barnaby, Frank, ed. *World Armaments and Disarmament: SIPRI Yearbook 1980.* Stockholm International Peace Research Institute. New York: Crane, Russak & Company, Inc., 1980.

Collins, John M. *U.S.–Soviet Military Balance: Concepts and Capabilities 1960–1980.* New York: McGraw-Hill, 1980.

The Defense Monitor. Center for Defense Information, 122 Maryland Avenue, N.E., Washington, D.C., 20002.

The Effects of Nuclear War. Op. cit.

Georgescu-Roegen, Nicholas. *The Entropy Law and the Economic Process.* Cambridge, Massachusetts: Harvard University Press, 1971.

Howard, Ted and Rifkin, Jeremy. *Entropy: A New World View.* New York: Viking, 1980.

Howe, Russell Warren. *Weapons.* New York: Doubleday, 1980.

Marshall, Eliot. "Iraqi Nuclear Program Halted by Bombing." *Science,* October 31, 1980.

"Nato and the Warsaw Pact." *The Economist,* August 9, 1980.

Podhoretz, Norman. *The Present Danger.* New York: Simon and Schuster, 1980.

Ramberg, Bennett. *Destruction of Nuclear Energy Facilities in War: The Problem and the Implications.* Lexington, Massachusetts: Lexington Books, 1980.

U.S. Strategic Bombing Survey. "The Effects of Strategic Bombing on the

German War Economy." Washington, D.C.: U.S. Government Printing Office, 1947.

U.S. Strategic Bombing Survey (Pacific). "The Electric Power Industry of Japan." Electric Power Division, December 3, 1945. Washington, D.C.: U.S. Government Printing Office, 1945

CHAPTER 3

Brown, William M. *Postattack Recovery Strategies—1980 Report to FEMA.* Croton-on-Hudson, N.Y.: Hudson Institute, November, 1980.

Calder, Nigel. *Nuclear Nightmares: An Investigation into Possible Wars.* New York: The Viking Press, 1979. 168 p.

Civil Defense for the 1980s—Current Issues. Washington, D.C.: Civil Defense Preparedness Agency, July 13, 1979. Unpublished.

Cooper, Ann. "Nuclear Evacuation: Don't Bet on It." *Washington Post,* July 20, 1980.

DCPA Attack Environment Manual. Defense Civil Preparedness Agency, Department of Defense. Washington, D.C.: U.S. Government Printing Office, 1978.

Hamilton, Alexander; Madison, James; and Jay, John. *The Federalist Papers.* New York: The New American Library, 1961.

Kahn, Herman. *On Thermonuclear War.* Princeton, N.J.: Princeton University Press, 1960.

Kaplan, Frank M. "The Soviet Civil Defense Myth—Part 2." *Bulletin of the Atomic Scientists,* April 1978.

Laurie, Peter. *Beneath the City Streets: A Private Inquiry into Government Preparation for National Emergency.* London: Panther Books, Granada Publishing, Ltd., 1979.

Lens, Sidney. *The Day Before Doomsday.* New York: Doubleday & Company, Inc., 1977.

Rapoport, Roger. *The Great American Bomb Machine.* New York: E. P. Dutton & Company, Inc., 1971.

State of California Emergency Program, Part 4, War Emergency Plan. California Office of Emergency Services (OES), State of California, 1970.

Yegorov, P. T.; Shlyakhov, I. A.; and Alabin, N. I. *Civil Defense: A Soviet View* [Grazhdanskaya Oborona]. Washington, D.C.: U.S. Government Printing Office, 1970.

CHAPTER 4

Fesharaki, Fereidun. *Global Petroleum Supplies in the 1980s,* February, 1980. "Availability of Strategic Materials," *Aviation Week and Space Technology,* May 5, 1980.

Georgescu-Roegen, Nicholas. *op. cit.*

Holden, Constance. "Energy, Security and War", *Science,* vol. 211, February 13, 1981.

Levy, Walter J. "Oil and the Decline of the West." *Foreign Affairs,* vol. 58, no. 5, Summer, 1980.

Ramberg. *Op. cit.*

Scott, Justin. *Shipkiller.* New York: Fawcett Books, 1979.

CHAPTER 5

Bethell, Thomas N. "How to Keep It Going: Synfuels." *Washington Monthly,* October 1980.
Cameron, Juan. "Washington's Ill-Starred Efforts to Stash Crude." *Fortune,* September 8, 1980.
Deese David A. and Nye, Joseph S. *Energy and Security.* Cambridge, Massachusetts: Ballinger Publishing Company, 1981.
Donnelly, Dr. Warren. *Nuclear Proliferation Factbook.* Congressional Research Service, U.S. Library of Congress. Prepared for the U.S. Senate Committee on Governmental Affairs and the U.S. House of Representatives Committee on Foreign Affairs. Washington, D.C.: U.S. Government Printing Office, September, 1980.
Emshwiller, John R. "Some Investors Shun Nuclear—Powered Utilities, Jeopardizing Funds to Build New Atomic Plants." *Wall Street Journal,* November 20, 1980.
Lilienthal, David E. *Atomic Energy: A New Start.* New York: Harper & Row, 1980.
Strategic Petroleum Reserve—Annual Report. U.S. Department of Energy, February 16, 1980. Washington D.C.: U.S. Government Printing Office, 1980.

CHAPTER 6

Adams, Henry. "A Letter to American Teachers of History." In *The Degradation of the Democratic Dogma.* New York: Capricorn Books, 1958.

CHAPTER 7

Brady, Jerry and Zimbler, B. "Conservation, Not Synfuels, Will Cut Oil Imports." *Energy User News,* August 18, 1980.
Clark, Wilson. "Rolls-Royce: The Volkswagen of the Nuclear Power Industry." *Science 80,* December, 1980.
"Decentralized Electricity and Cogeneration Options." Energy Committee of the Aspen Institute for Humanistic Studies, Second Annual R&D Workshop, Aspen, Co., July 13–17, 1979.
"Downsizing Nuclear Plants." *Business Week,* November 10, 1980.
Dubin, Fred and Long, Chalmers. *Energy Conservation Standards.* New York: McGraw-Hill, 1978.
Energy in Transition: 1985–2010. Committee on Nuclear and Alternative Energy Systems (CONAES). National Academy of Sciences: San Francisco: W.H. Freeman and Co., 1980.
Ford, Andrew. "A New Look at Small Power Plants, Is Smaller Better?" *Environment,* vol. 22, no. 2, March, 1980.
Ford Andrew and Yabroff, Irving W. "Defending Against Uncertainty in the Electric Utility Industry." *Energy Systems and Policy,* 1979.
Gibbons John H. and Chandler, William U. *Energy: The Conservation Revolution.* New York: Plenum Press, 1981.
O'Neal, Dennis; Carney, Janet; and Hirst, Eric. *Regional Analysis of Residential Water Heating Options: Energy Use and Economics.* Oak

Ridge National Laboratory, Oak Ridge, Tennessee, October, 1978. Springfield, Va.: National Technical Information Service, U.S. Department of Commerce, 1978.

Ross Marc H. and Williams, Robert H. *Our Energy: Regaining Control.* New York: McGraw-Hill, 1981.

Sant, Roger. *The Least-Cost Energy Strategy: Minimizing Consumer Costs Through Competition.* Arlington, Va.: The Energy Productivity Center, Mellon Institute, 1979.

Schuyten, Peter J. "American Industry Faces Huge Capital Investments to Increase Efficiency." *New York Times,* January 11, 1981

"Six Governors on the Auto Industry." Letters to the Editor. *Washington Post,* March 7, 1981.

Stobaugh, Robert and Yergin, Daniel. "Energy: An Emergency Telescoped." *Foreign Affairs,* vol. 58.3, 1980.

Wright, Patrick. *On a Clear Day You Can See General Motors.* New York: Basic Books, 1980.

CHAPTER 8

Asbury, J. G.; Geise R. F.; and Mueller, R. O. "Electric Heat: The Right Price at the Right Time." *Technology Review,* December/January, 1980.

Calsetta, Alfred B. *Electric Utility Load Research Data Processing System.* Tennessee Valley Authority, October, 1979.

Cichetti, C. and Reinberg, J. "Electricity and Natural Gas Rate Issues." *Annual Review of Energy,* 1979.

Considine, Douglas M., ed. *Energy Technology Handbook.* New York: McGraw-Hill, 1977.

Kaufman, Alvin. "Will the Lights Go On in 1990?" Congressional Research Service, Library of Congress. Washington, D.C.: U.S. Government Printing Office, August, 1980.

1980 Conservation and Load Management Program. Southern California Edison Co. Rosemead, CA., December, 1979.

Post, Richard F. and Post, Stephen F. "Fly-wheels." *Scientific American,* December, 1973.

"A Proven Way to Save Energy: Cogeneration." Washington, D.C.: Manufacturing Chemists Association, 1978.

"Putting Baseload to Work on the Night Shift." *EPRI Journal,* vol. 5, no. 3, April, 1980.

Ross & Williams. *Op. cit.*

Smith, Tim. "EMS Market Growth Spurred by Interest in Wireless Units." *Energy User News,* February 16, 1981.

Williams, Robert H. "Industrial Cogeneration." *Annual Review of Energy,* vol. 3, Palo Alto, CA. Annual Review, Inc., 1978.

CHAPTER 9

Anderson, Bruce. *Solar Energy: Fundamentals in Building Design.* New York: McGraw-Hill Book Company, 1977.

Application of Solar Technology to Today's Energy Needs, vol. 2. Office of Technology Assessment. Washington, D.C.: U.S. Government Printing Office, September 1978.

Flavin, Christopher. *Energy and Architecture: The Solar and Conservation Potential.* Worldwatch Paper 40. Washington, D.C.: Worldwatch Institute, November, 1980.

Clark, Wilson. *Energy for Survival.* New York: Anchor Press/Doubleday, 1974.

Fire of Life: The Smithsonian Book of the Sun. New York: Smithsonian Exposition Books/W. W. Norton & Company, 1981.

The Future of Solar Electricity, 1980–2000: Developments in Photovoltaics. Gaithersburg, Md.: Monegon, Ltd., January 1980.

Lovins, Amory. *Soft Energy Paths: Toward a Durable Peace.* Cambridge, Massachusetts: Friends of the Earth/Ballinger Press, 1977.

Renewable Ocean Energy Resources, part I. Office of Technology Assessment. Washington, D.C.: U.S. Government Printing Office, May, 1978.

Solar Engineering Magazine, April, 1980. Articles on solar ponds in Israel; Miamisburg, Ohio; and Salton Sea.

Stein, Richard. *Architecture and Energy.* New York: Anchor Press/Doubleday, 1977.

CHAPTER 10

"Alternative Budget Proposals for the Environment: 1981, 1982." Washington, D.C.: Environmental Defense Fund, et al., March, 1981.

An Analysis of Federal Incentives Used to Stimulate Energy Production, revised. Battelle Memorial Institute. Richland, Washington: Pacific Northwest Laboratory, 1978.

"Annual Average Wind Power (Watts/M²) at 50M." Richland, Washington: Pacific Northwest Laboratory, 1980.

Assessment of Technology for Local Development. Office of Technology Assessment. Washington, D.C.: U.S. Government Printing Office, 1981.

Azarin, Beverly. "In the Wake of the Flying Cloud." *Science 81,* March, 1981.

Clark, Wilson. *Energy for Survival. Op. cit.*

Direct Utilization of Geothermal Energy: A Technical Handbook. Geothermal Resources Council, Report no. 7, 1979.

Energy from Biological Processes. Office of Technology Assessment. Washington, D.C.: U.S. Government Printing Office, August, 1980.

Fuel Alcohol: An Energy Alternative for the 1980s. Final Report, U.S. National Alcohol Fuels Commission. Washington, D.C.: U.S. Government Printing Office, 1981.

"Global Future: Time to Act." Council on Environmental Quality. The White House. Washington, D.C.: U.S. Government Printing Office, 1981.

"Going with the Wind." *EPRI Journal,* March, 1980.

Kruger, Paul. "Geothermal Energy." *Annual Review of Energy.* Palo Alto, Ca.: Annual Review, Inc., 1976.

Large Wind Turbine Generator Performance Assessment, Technology Status, Report no. 1. Boston Massachusetts: Arthur D. Little, Inc., January, 1980.

Lerner, Dr. James. California Energy Commission, 1111 Howe Avenue, Sacramento, California. Primary contributor on wind energy to this section; excellent resource base in all aspects of wind power.

Moorer, Admiral Thomas H. "Written Testimony Before the Subcommittee on Energy, Nuclear Proliferation, and Government Processes of the Committee on Governmental Affairs," March 24, 1981.

Odum, Howard T. *Environment, Power and Society.* New York: Wiley, 1971.

Peattie, Donald Culross. *Flowering Earth.* New York: Viking, 1965.

"Wind Energy Development." Windfarms, Inc., 639 Front Street, San Francisco, CA., 94111. (1980)

EPILOGUE

Beres, Louis Rene. *Apocalypse: Nuclear Catastrophe in World Politics.* Chicago: University of Chicago Press, 1980.

Brandt, Willy. *North-South: A Program for Survival.* Boston: MIT Press, 1980.

Bush, Vannevar. *Modern Arms and Free Men.* New York: Simon and Schuster, 1949.

Drell, Sidney. "Arms Control: Is There Still Hope?" *Daedalus,* fall, 1980.

Federalist Papers. Op cit.

Garwin, Richard L. Letter to U.S. Representative John Seiberling, February 4, 1980.

Greene, Wade. "Rethinking the Unthinkable." *New York Times Magazine,* March 15, 1981.

Kincade, William. *Daedelus,* fall, 1980.

Kohr, Leopold. *The Breakdown of Nations* (reprint). New York: E.P. Dutton, 1978.

Lodal, Jan. "Deterrence and Nuclear Strategy." *Daedalus,* fall, 1980.

Mech, David. *The Wolf.* New York: Natural History Press, 1970.

Sivard, Ruth L. *World Military and Social Expenditures—1980.* World Priorities, Box 1003, Leesburg, Virginia 22075.

APPENDIX

Dispersed, Decentralized and Renewable Energy Sources: Alternatives to National Vulnerability and War. Energy and Defense Project. (California Academy of Sciences) Washington, D.C.: Environmental Policy Institute, 1981. Copies of the technical report are available from the Environmental Policy Institute, 317 Pennsylvania Avenue, S.E., Washington, D.C. 20003.

Index